INSTRUCTIONAL SYSTEMS DEVELOPMENT IN LARGE ORGANIZATIONS

WALLACE HANNUM
University of North Carolina at Chapel Hill

CAROL HANSEN

EDUCATIONAL TECHNOLOGY PUBLICATIONS
ENGLEWOOD CLIFFS, NEW JERSEY 07632

Library of Congress Cataloging-in-Publication Data

Hannum, Wallace H.
    Instructional systems development in large organizations/Wallace Hannum, Carol Hansen.
      p.   cm.
    Bibliography: p.
    Includes index.
    ISBN 0-87778-204-0
    1. Employees, Training of. 2. Instructional systems—Design.
I. Hansen, Carol.   II. Title.
HF5549.5.T7H296   1989   88-35237
658.3'12404—dc19   CIP

Copyright © 1989 Educational Technology Publications, Inc., Englewood Cliffs, New Jersey 07632.

All rights reserved. No part of this book may be reproduced or transmitted, in any form or by any means, electronic or mechanical, including photocopying, recording, or by information storage and retrieval system, without permission in writing from the Publisher.

Printed in the United States of America.

Library of Congress Catalog Card Number:
88-35237.

International Standard Book Number:
0-87778-204-0.

First Printing: January, 1989.

INSTRUCTIONAL SYSTEMS DEVELOPMENT IN LARGE ORGANIZATIONS

Preface

This book grew from a discussion one snowy afternoon in Williamsburg about courses we were teaching at separate universities for persons who had or were about to have responsibilities for developing training, usually in large organizations. Our desire was to focus on the macro level of instructional systems, to teach the "big picture," rather than teaching how to write lesson objectives and create instructional segments. There are many excellent books that deal with the specifics of designing individual lessons but few that deal with the design and development of large scale instructional systems. Those books that focused on instructional systems approached it from a more mechanistic, reductionistic stance than we wanted to do in our courses. Developing instructional systems is a complex, high level undertaking involving many decisions requiring informed judgment with constant trade-offs and adjustments to be made. A simple 1-2-3-4 "paint by the numbers" approach doesn't capture the realities of developing and operating instructional systems in most organizations.

Since neither of us found a book we could use in our courses, we used selected readings in lieu of a textbook. Students found the readings helpful but nevertheless disjointed and asked for a textbook. Through a series of conversations, we identified the content for a basic textbook on Instructional Systems Development and the approach it should take. With the support, and patience, of Larry Lipsitz, our publisher at Educational Technology Publications, we have developed such a book, describing what is known about developing instructional systems in large organizations.

We believe this book will prove useful as the text for courses in Instructional Systems Development as well as a reference guide to assist persons with responsibility for developing and managing large scale education and training programs. Our intent is to present information about the process one goes through in developing instructional systems rather than provide a fixed step by step sequence of specific actions. We believe Instructional Systems Development must be more flexible than a linear model appears to be so it can be adapted to organizational realities. Developing instructional systems requires judgment about what to do and when to do it. Instructional Systems Development is professional work, not a clerical task to be accomplished by the uninformed.

There are increased pressures on most organizations in our society to increase their performance through improved education and training. There are pressures on businesses to improve worker productivity, on educational systems to increase students' achievement, on governmental agencies to improve employees' performance, and on the military services to operate more efficiently in an uncertain world. While the value of education and training has been demonstrated time and time again, most large organizations fail to take advantage of what is known about designing and developing educational and training programs. We believe that education and training can be enhanced significantly through systematic application of the Instructional Systems Development principles found in this book.

We are especially pleased to include three case studies that illustrate the application of an Instructional Systems Development approach within large organizations. Robert Branson has documented the use of Instructional Systems Development in the Army and Navy to solve the problem of basic skills training. Dr. Branson is a professor at Florida State University and has spent several years working with large organizations, especially the military services, in developing educational and training programs. He was the primary developer of the Interservice Procedures for Instructional Systems Development model that defined how military education and training was to be done. He continues to be a frequent contributor to the literature on Instructional Systems Development.

Preface

Martha Brooke has described the use of Instructional Systems Development in a large scale management training program in the Internal Revenue Service. Dr. Brooke is an education specialist at the Internal Revenue Service with several years experience applying Instructional Systems Development principles to training programs. Richard Duggins has described the development of programs to train service personnel within IBM. A long time IBM employee, Mr. Duggins has been an instructor, a course developer, and most recently a manager of course development. An engineer by training, Mr. Duggins worked in the training environment at IBM for 17 years.

Instructional Systems Development is a way of approaching the tasks of providing education and training. It is not a mystical panacea guaranteed to solve all problems in any situation. But when conscientiously applied, Instructional Systems Development principles can, and should, result in more effective and efficient education and training programs. We hope this book can assist others in using this powerful approach in education and training.

W.H.
C.H.

Table of Contents

Preface v
Chapter 1. Introduction to Instructional Systems
    Development. 3
Chapter 2. Benefits and Requirements for Instructional
    Systems Development. 11
Chapter 3. The ISD Model. 25
Chapter 4. Needs Assessment 39
Chapter 5. Performance Analysis. 67
Chapter 6. Task Analysis. 81
Chapter 7. Goals and Objectives 93
Chapter 8. Analysis of Intended Participants. 115
Chapter 9. Organizing and Sequencing Instructional
    Content. 123
Chapter 10. Methods and Media Selection. 139
Chapter 11. Project Management and Consulting 157
Chapter 12. Implementing Planned Change. 171
Chapter 13. Evaluation in Instructional Systems
    Development. 191
Chapter 14. Future Directions for Instructional
    Systems Development. 207
Appendix A. Large Scale ISD Programs. Two Case Studies
    in the Military Services. By Robert K. Branson. 225
Appendix B. Instructional Systems Development in a Large
    Governmental Agency. By Martha Brooke 255
Appendix C. Instructional Systems Development in a
    Large Computer Company. By Richard Duggins. 289
References. 303
Index. 309

INSTRUCTIONAL SYSTEMS DEVELOPMENT IN LARGE ORGANIZATIONS

Chapter 1
Introduction to Instructional Systems Development

Almost daily we read and hear that in the United States we are losing our preeminent position among nations in the world. Our businesses and industries are reported to have lost their competitive edge. We face continuing trade deficits as our products are less attractive in the world market while we buy a record number of foreign goods. We hear of domestic plants closing as companies go off-shore for labor. Foreign goods, particular consumer goods, are thought to be of better quality than items produced in America. While we were once leaders in world trade and in standards of living our citizens enjoyed, we have been surpassed by some other industrialized nations. In a similar manner, our educational system, once the envy of the world, has been shown to fare poorly when compared with those of many other nations. Despite large expenditures of public monies on education, our students don't seem to be learning as much as students in other nations.

As we enter the 1990's the United States does not seem poised to prosper; our citizens do not look towards an increased quality of life. Undoubtedly, there are many factors responsible for the much touted decline experienced in America. Certainly economic factors play a large role, as do the governmental policies of our trading partners. Yet most analyses point to the abilities of our workforce, our citizens, as perhaps the most important factor underlying the problems we are currently facing. That is, our children are no longer excelling in school when compared to children in other countries; our workers likewise are no longer excelling in their jobs when compared with workers in other nations.

The Challenge

We don't propose yet another rehash of our economic woes nor do we advocate yet another solution to these problems. We wish to indicate that many authors and analysts have identified what perhaps is our nation's greatest need—the need for better educated citizens and better trained workers. This is a need that grows as our economy continues the transformation to a third wave or information economy. Traditional ways of meeting education and training needs are not proving effective for this challenge. Rather we require a new "technology" for education and training, a technology that allows us to surpass what has been possible with traditional approaches to education and training. This book attempts to present such a technology. We are not striving to develop a new technology, for we believe that appropriate educational and training know-how already exists. The intent of this book is to compile that knowledge and present it in a form that is understandable to persons who have responsibility for the development and management of education and training programs, and particularly in large organizations, such as corporations, government agencies, military services, etc.—wherever there are many individuals to train and educate under the aegis of large numbers of instructional as well as managerial and design and development personnel. In such entities, many people must work together toward the common goal of designing and developing a system to be used in the training and education of hundreds and often many thousands of employees, trainees, students, customers, and sometimes even consumers. Instructional Systems Development, as described in this book, offers organizational managers methods and means for reaching the goal of achieving the best possible training and education for their subordinates, colleagues, patrons, etc., whatever the designation of the student or trainee might be in a specific situation. It is a complex process, requiring the timely interaction and cooperation of numerous levels of management and personnel, often situated in many parts of the country and even throughout the world.

Instructional Systems Development

Whether you have, or will have, the responsibility for training equipment maintenance workers, lawyers, computer programmers, clerical workers, high school teachers, or soldiers there are some procedures for designing, developing, and offering training programs that will more likely result in success than traditional methods. These procedures as a whole or in parts have been described by many terms—educational technology, instructional technology, instructional design, instructional development, instructional systems technology, and instructional systems development. Because we view these procedures as the basis for developing instructional systems, we will use the term instructional systems development (ISD) throughout this book.

Training programs. The focus in this book will be on developing education and training *programs* and not on developing individual lessons. There are many excellent sources of information for the reader who is interested in developing individual lessons, including Gagne, Briggs, and Wager (1988), Dick and Carey (1985), Fleming and Levie (1978), and Kemp and Dalton (1985). Since this book is focused on education and training programs rather than lessons, the selection and coverage of topics reflects the concerns of persons who have responsibilities for developing and managing such programs. Thus, this book will contain guidance for planning a program, not for making instructional material. It is a book for managers rather than technicians.

Origins. The approach presented in this book has evolved from many sources over the past three decades. Certainly all ISD efforts owe some debt to the military services. Many of the procedures used in instructional systems development had their origins in the military. Persons in several universities and other research organizations have conducted research on which ISD stands. These persons often had their own training in experimental psychology, educational psychology, or communications.

Some of the early applied work in ISD was done by educators, particularly those with an interest in audio-visual instruction, and military trainers who had widely varying backgrounds. There are many contributions to ISD from training programs in several companies. Other aspects of ISD have emerged from work in public education and from education in developing nations. In short, there are many places from which ISD began. There are many ideas about improvements in education and training programs that were developed separately and now are being assembled into a more coherent system.

Applications of ISD. Principles of instructional systems development seem to be applied more often in those situations where expectations about the outcomes of the education or training can be clearly established and the organization providing the education or training is held accountable for producing results. ISD projects flourish in organizations that place *an economic value on the time of the learners or students*. Thus, it is more common to see ISD principles followed in a company that pays employees during the training than to find ISD employed in public schools.

As you will see in the second chapter, a common finding from research on ISD is that the learner usually requires less time for instruction when ISD is used than in traditional instruction. So when the learners' time has a greater economic value, ISD will be more valued. For example, if the application of ISD in a business could reduce a particular training program for a class of employees from 30 hours to 20 hours, the company would gain 10 hours from each employee to be spent producing a product or providing a service for the company. In short, the employees could spend more time *earning* money for the company rather than *costing* the company money. If ISD could be used by a school in the rural area of a developing country, then the children could spend less time in school and more time working on the family farm to help support their families. Indeed, in many developing countries the labor of children is often essential if their families are to maintain even a subsistence level of farming. Thus, the children can't attend school five days a week for nine months. In these situations an alternative approach to education that is

Introduction to Instructional Systems Development

consistent with the constraints of the environment must be used; otherwise most students will become dropouts very early. This book, however, concentrates primarily on the non-school sectors of the world economy, since most schools simply do not function as large organizations. Rather, the vast majority of instructional decisions remain isolated in the hands of individual classroom teachers, who do not function as part of a truly systematic framework.

ISD is becoming more commonplace in business and industrial training because each employee must be able to perform certain tasks or functions to contribute to his or her company's success. Much of a company's success depends on the quality of its employees' work. Better trained employees are more successful in (1) getting their work done, (2) in an acceptable fashion, (3) within a reasonable time period.

When companies fund training programs they do so because they believe that the training program will contribute to "bottom line" effectiveness, to profitability. Perhaps a company's most precious assets are the knowledge and skills of its employees. By increasing these knowledge and skills, a company often can increase its profits. In short, corporate training is an investment, and like any other investment it is expected to have a return to the company. When training programs are effective in teaching essential knowledge and skills to employees, and do so efficiently, the investment in those training programs can be among the best investments that a company might make. The same applies to government employees and military personnel, although the end-results in these instances are measured in organizational effectiveness rather than monetary profits. Obviously, the sooner an agency clerk gets on the job, or the sooner a soldier takes to the field, the better it is for the proper functioning organizational entities involved.

Strategic planning. There is another aspect of training programs that is important in business and industry. In order to prosper, perhaps even survive, in competitive markets companies must produce goods or provide services that are in demand and seen as superior to those of their competitors. Guiding a company that is to remain successful requires long-range planning as well as careful daily management. The ever-changing marketplace requires new products and new services. The company that identifies these products and services and provides them is more likely to prosper. However, each time a company makes a change in its product line, the employees will likely have to change how they do their work. Additional training is required. If companies are to be successful, then training must be a part of their strategic planning, on a continuing basis. In different contexts, government, military, and other large non-business organizations require continuing training and education efforts, despite the fact that their end goals are matters other than profits for shareholders.

Many of the companies most cited for excellence are also companies that have ambitious employee training programs to support their plans and efforts. These are environments in which ISD approaches are most often found. Many companies even have campuses to conduct their education and training. Some large firms offer complete curricula designed to take new personnel from entry levels through top level management. Many firms prefer to promote from within for this reason. Their training includes more than just procedures required for their jobs; the training also integrates them into a *culture*. Firms want employees to learn *their* procedures and *their* culture. Appropriately conceived training programs support an organization's overall goals by providing the employees the necessary knowledge and skill to successfully perform their jobs. Training is becoming more important to organizations, and training is receiving more support. Training programs are also coming under more pressure to produce, to effectively and efficiently train employees. It is our belief that instructional systems development procedures offer the best hope for successful education and training programs.

Organization of This Book

The chapters in this book cover various topics associated with instructional systems development. Chapter 2 provides a discussion of what benefits you can expect to follow the successful application of ISD in education and training settings. This chapter also identifies some of the requirements for ISD. While ISD has been demonstrated to work in a variety of circumstances, there is no magic in ISD; it can just as easily fail if certain requirements are not attended to. Chapter 3 presents a generic model for ISD by identifying steps that should be followed in developing instructional systems. This chapter includes a comprehensive listing of steps with a description of each step. Of course, "doing" ISD involves more than blindly following such steps as a recipe. The steps contained in this, or any other, model are guidelines to be considered when attempting to implement ISD in an education or training environment. Adjustments or modifications to this model are inevitable but should be done while keeping in spirit with the model (Hannum, 1984). Chapter 4 discusses the use of needs assessment techniques as the way to begin any ISD effort. It is through a complete needs assessment that the appropriate basis for any education or training program should be established (Kaufman, 1985, 1986). By beginning with a needs assessment, instructional systems development efforts establish a firm direction so that the goals pursued in the education and training program are indeed responsive to the environment of the program. Through needs assessments, appropriate goals can be identified, and then the training effort can be focused on meeting these goals. Chapter 5 describes a technique called performance analysis, which is used to determine the most probable cause for the identified needs and to suggest possible solutions to alleviate or reduce these needs. Chapter 6 describes procedures for identifying the essential skills a person must possess to perform the various tasks that make up a certain job. These job analysis procedures are used when the need is shown to arise because employees lack the necessary skill or knowledge to perform their jobs successfully, or because new employees must be trained for these jobs. Chapters 4, 5, and 6 constitute what is often termed "front-end analysis," since they are the initial steps, or front-end, of an ISD approach. Chapter 7 describes procedures for developing goals and, in turn, objectives based on the results of the front-end analysis. Chapter 8 includes a description of how to consider the audience, or recipients, of the education or training when developing the program. Successful education or training requires that the instruction is appropriate for the people receiving the instruction (Ausubel *et al.*, 1978); it is essential to consider the audience when planning programs. Chapter 9 presents techniques for organizing and sequencing instructional content to support education and training programs. The instructional or training goals must be organized into courses and the courses must be further organized into lessons to make a coherent program. Effective organization and sequencing of the instructional content is essential if the programs are to be successful. Chapter 10 discusses the selection and use of instructional methods and media to deliver the instructional content. Much of the front-end analysis is done to ensure that the resulting training will be effective, that is, that the training will focus on important goals and objectives. The selection of instructional methods and media ensure that the training will be as efficient as possible. Chapter 11 describes some procedures for the management and implementation of ISD projects. As with most efforts, ISD projects must be carefully managed if they are to be successful. This chapter presents some concerns about the management of ISD projects. Chapter 12 discusses the process of introducing a planned change within an organization. Instructional systems development efforts result in the implementation of a new order of things with regard to education and training. Research by Berman and McLaughlin (1978) indicates that *how* an educational change is introduced is more important than the specific change itself in determining success. It is not sufficient to develop a better training program. Care must be taken in implementing even the very best program, or it is likely to fail. Chapter 13 describes the role of evaluation in instructional systems development and various approaches to evaluation. Chapter 14 summarizes instructional systems development and looks towards the future. Three case studies (Appendicies A, B, and C) illustrate the application of ISD principles in three different settings. Appendix A reports on the use of ISD within the Army and Navy for basic skills training. Appendix B reports on the use of ISD for internal training within a large governmental agency, The Internal Revenue Service. Finally, Appendix C reports on the use of ISD in a large corporate training setting.

Chapter 2
Benefits and Requirements for Instructional Systems Development

The decision to undertake an instructional systems development project should be based upon some reasonable expectations regarding the probable outcomes or benefits from such an endeavor. The initial costs of an ISD project are often higher than traditional instruction, in which individual instructors are placed into classrooms of students or trainees and left to "teach." Should the probable benefits be scant and the costs of undertaking a project high, then the use of instructional systems development would seem unwise. However, if the real long-term *value* of the expected outcomes exceeds the associated costs, then an instructional systems development project would seem a wise investment. Indeed, ISD is an investment for a large organization, but we feel that, when properly done, instructional systems development can be one of the better investments an organization might make.

Reasonable Expectations from ISD

In this chapter we will explore some of the benefits that will likely occur as a result of instructional systems development efforts, and we address some of the requirements necessary for ISD. While we are excited about the possibilities ISD can generate within an organization, we are mindful of its limitations and the conditions under which ISD doesn't seem to flourish. ISD is not a cure-all that can remedy all that ails an organization. ISD does not always solve every training problem; neither is it guaranteed to be equally effective in all situations. The purpose of this chapter is to assist in identifying reasonable expectations from ISD, requirements for successful implementation, and limitations associated with ISD.

Research on ISD

Course development. First, let's explore the benefits that have accompanied instructional systems development projects. One of the earliest reports of the effectiveness of ISD emerged from a course for training personnel responsible for maintaining long distance communications circuits and equipment at AT&T (Mager, 1977). As is so often the case, their personnel had been trained in a rather traditional lecture class on electronic fundamentals. There were indications that the course did not prepare the students adequately for their jobs. The supervisors of these students also failed to see that the course helped prepare them for their job assignments. Thus AT&T decided to engage in a training development project.

Analysis. Since they had already determined the need for a training program, the first step undertaken in this effort was to conduct a task analysis, i.e., to systematically examine what tasks are routinely performed by persons in the jobs for which the students were being trained. Detailed descriptions of the tasks performed by workers on the job were collected and analyzed. These descriptions included information about: (1) the conditions at the time the specific activity took place, (2) the activity itself, and (3) feedback resulting from the activity. Data were also collected regarding how often the workers did each specific task, which job assignments included the task, and the percentage of workers who would have these job assignments. This information about the tasks performed was then used as the basis for identification of instructional content. That is, the course developers used this task analysis data to determine what to include in the course on fundamentals. In such a fashion the selection of course content was based upon empirical data rather than someone's opinion about what content to include. Of the 280 separate activities identified in the task analysis, a restricted set of 30 activities accounted for half the tasks they observed being performed. This limited set of 30 activities was considered the basic set of tasks the trainees would be expected to master.

Benefits and Requirements

Using data to develop courses. The data from the task analysis were used as the basis for developing the course material. Instruction was planned for those tasks that constituted the 30 basic activities. Conversely, there was to be no instruction on any activities that were not included in the basic 30 tasks. By using task analysis data, the course developers were able to "line up" their course content with the actual job tasks.

Evaluation and revision. The resulting course went through four tryouts. In the first tryout the training methods were similar to the traditional course, but the content of the tryout version reflected the task analysis, old methods, and new objectives. In the second tryout the course was individualized using self-study modules and self-pacing progress. The third and fourth tryouts of the course consisted of evaluation and revision of the materials based upon performance data from students. Data from these four tryouts are compared with data from the traditional course.

Results. As can be seen in Table 2.1, there were some remarkable findings from this project. The amount of time spent in training was markedly reduced from 45 days in the traditional course to nine days after the fourth tryout. The performance of students increased from 80 percent in the first tryout of the new course to 92 percent in the third and fourth tryouts. Unfortunately, performance data were not collected in the traditional course, but the performance levels in any case had not been judged satisfactory; this is what led to the project being undertaken in the first place. As a result of the training development process, the achievement of the students was increased while training time was reduced. The new course was both more effective and more efficient, in other words, a success.

Costs and savings. There were some costs associated with this project for the task analysis and materials development. The task analysis effort required approximately $200,000 and the materials development another $150,000 for a total project cost of about $350,000. The savings, however, were much greater. The savings in the initial year alone were $2,000,000, more than offsetting the project's costs. These savings were the result of less employee time spent in training and, therefore, more time spent providing services that earned the company money. The savings over the first five years were estimated to be over $37,000,000! Quite a return on investment, by any standard.

Table 2.1 Improvements from Using ISD.

Cycle	Technique	Time in days	Score
0	Lecture	45	?
1	Select topics from task analysis	20	80
2	Individualize materials & rate	13	91
3	Formative evaluation & revision	9	92
4		9	92

ISD benefits. While these data are dramatic, they represent typical results from ISD efforts. It is reasonable to expect greater performance from the students and to achieve this performance in considerably less training time. The reduction in training time arises from the attention paid to the analysis phase of instructional systems development. Training time is reduced when the instruction is based upon task analysis data, and *non-essential content is eliminated*. Since training is provided for the essential tasks and since the training is evaluated and revised until the students actually master these tasks, higher levels of criterion performance usually result. These are the primary benefits of using an instructional systems development, approach: (1) increased achievement levels, (2) decreased training time, resulting in more time on the job, (3) decreased instructor dependence with less presentation variance, and (4) greater performance accountability.

Program benefits. While these benefits can and do accrue when ISD is used in an individual course of instruction, the implications are more dramatic when ISD is applied to training programs composed of many learners participating in several different courses in various locations over a period of years. When all the courses in a training program or a curriculum are approached systematically through the use of instructional systems development principles, the resulting savings of training time and costs are much greater that those from an isolated application of ISD. Further, these cost savings continue over the life of the training program, since each subsequent group receiving the training would require less time. In addition to these cost savings, there is another benefit that would likely arise when ISD is applied to training programs that is equally important to the economic benefits of ISD. There is a much greater likelihood that the learners would be able to perform at acceptable levels after completing the training. In short, the education or training program would enable most of the learners to master most of the instruction. Thus the "products" of the training programs, the students or employees, would be "acceptable" to their on-the-job supervisors when completing the training.

Materials development project. There are some other notable benefits from ISD that are illustrated in a case study reported by Markle (1977). This effort was a course development project for basic first aid training designed to meet American National Red Cross certification requirements. This project spent considerable time analyzing the subject matter and precisely defining the expected outcomes of the training. Specific test items were constructed to measure the outcomes expected from the students. Instructional films demonstrating first aid procedures and accompanying workbooks were developed. The training program then went through several evaluation and revision cycles. The intent within the project was to improve the instruction until it produced the required performance in the students. The project made extensive use of empirical data from these tryouts as a basis for revising the training program.

Results. The results from this project are summarized in Figure 2.1.

An analysis of this Figure 2.1 reveals some interesting findings about the results from instructional systems development projects. As with the other project cited in this chapter, the student showed higher achievement scores and decreased training time. You may note the initial increase in training time in the first ISD cycle. This reflects the addition of course content related to the criterion performance that was absent from the standard course. Later cycles in the development process reduced the training time as efficiencies were achieved. The average achievement score rose dramatically in the first ISD cycle, confirming that instructional systems development procedures can increase achievement when compared with traditional instruction.

ISD versus traditional instruction. There are some other interesting findings from this project in addition to these rather "standard" results. Not only was the average achievement higher after ISD but also the performance of the worse student, the lowest achievement attained, was considerably above the average achievement under the standard instruction. In essence, the worst student in the ISD group bettered the average performance in the standard group and was very close to equaling the highest performance attained in the standard group. A further glance at Figure 2.1 will

Benefits and Requirements 17

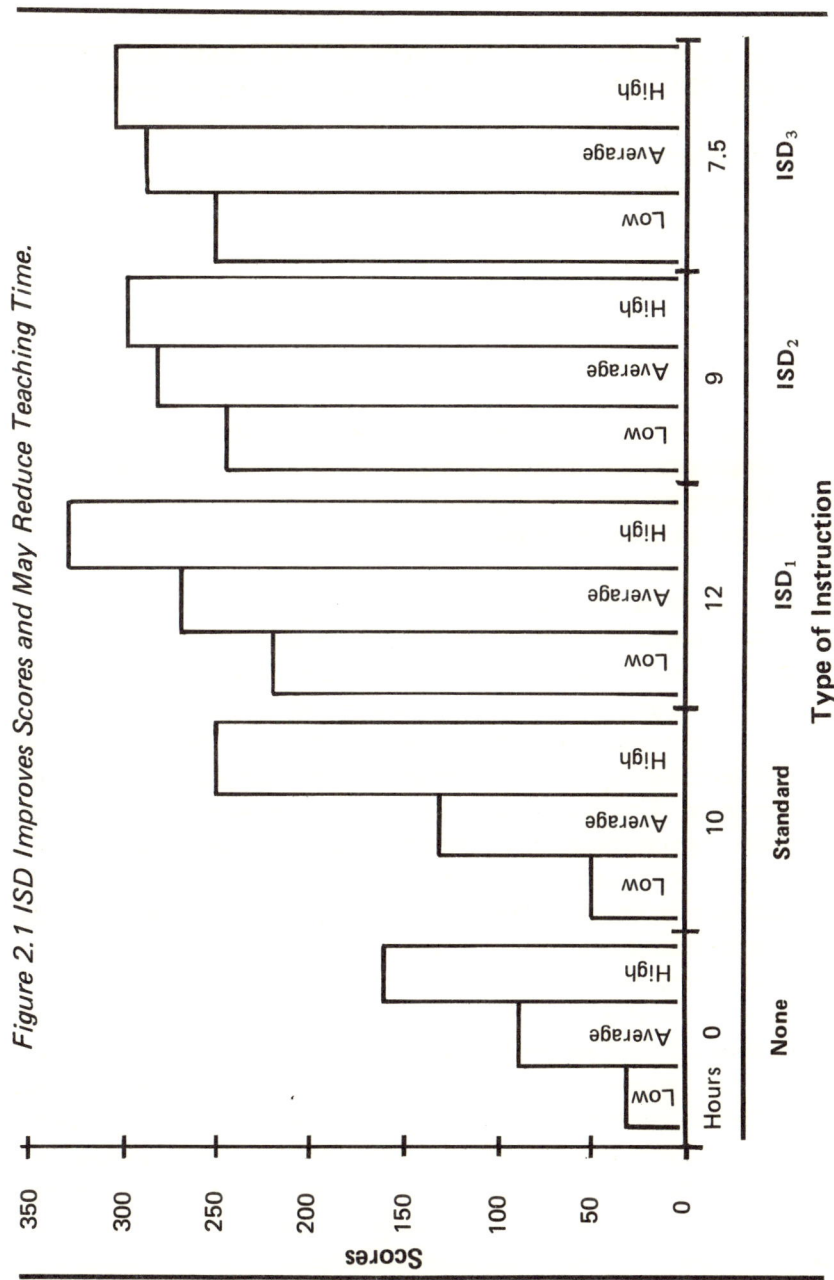

Figure 2.1 ISD Improves Scores and May Reduce Teaching Time.

also indicate that the spread or variance in performance among students commonly found in traditional instruction seems largely reduced, if not absent, in ISD training. Students completing ISD training are more alike in their achievement than students completing standard training courses; the ISD students' performance is more uniform. There is no so-called standard "curve" of good to poor results.

Expected Benefits from Using ISD

In summary, it seems reasonable to expect certain benefits as a result of instructional systems development efforts. There should be:

- increases in achievement
- greater mastery of tasks
- reduction in training time
- restriction in the range of achievement
- lower training costs
- more accountability for training results

Of course, not all of these benefits will accompany every ISD effort, no matter how it is conducted, but it is reasonable to expect similar results from carefully conducted projects. Some of these benefits may occur in some projects that don't fully implement instructional systems development principles but do use certain aspects of ISD. It is rare to find a project that is "all" ISD or that is "all" traditional instruction. Rather, there may be some blending of ISD principles into an otherwise traditional course or training program.

The benefits we have described are those that are likely to occur when ISD principles are used consistently. We do not expect that an educational or training program would use every ISD principle and no aspects of traditional instruction—this would be sheer folly. Rather, we believe that the spirit and intent of ISD must be followed. When this happens there are certain differences between the instructional systems program and traditional programs. Figure 2.2 presents a summary of these differences (Hannum and Briggs, 1982), which we believe are quite striking.

Benefits and Requirements

Figure 2.2 Comparison of Traditional and Systematic Instruction.

Component of Instruction	Traditional Instruction	Systematic Instruction
Setting Goals	—traditional curriculum —textbook —internal referent	—needs assessment —job analysis —external referent
Objectives	—stated in terms of global outcomes or teacher performance —same for all students	—from needs assessment/analysis —stated in terms of student performance —chosen with consideration for students' entering competencies
Students' Knowledge of Objectives	—not informed; must intuit from lectures and textbooks	—specifically informed in advance of learning
Entering Capability	—not attended to —all students have same objectives and materials/activities	—taken into account —differential assignment of objectives and materials/activities
Expected Achievement	—normal curve	—uniform high level
Mastery	—few students master most objectives —hit-and-miss pattern	—most students master most objectives
Grading and Promotion	—based on comparison with other students	—based on mastery of objectives
Remediation	—often not planned —no alteration of objectives or instructional means	—planned for students who need help —pursue other objectives —use alternate instructional means
Use of Tests	—assignment of grades	—monitor learner progress —determine mastery —diagnose difficulty —revise instruction
Study Time vs. Mastery	—time constant; degree of mastery varies	—mastery constant; time varies
Interpretation of Failure to Reach Mastery	—poor student	—need to improve instruction

Figure 2.2 (Continued)

Component of Instruction	Traditional Instruction	Systematic Instruction
Course Development	—materials selected first	—objectives stated first, then selection of materials
Instructional Media and Materials	—selected on basis of preference and availability —effectiveness not known	—selected on basis of objectives and characteristics of students —based on theory and research —must have demonstrated effectiveness
Sequence	—based on logic of content and outline of topics	—based on necessary prerequisites and principles of learning
Instructional Strategies	—"across the board" favorite —based on preference and familiarity	—selected to attain objectives —use of various strategies —based on theory and research
Evaluation	—often does not occur; rarely systematically planned —norm-referenced —data on inputs and processes	—systematically planned; routinely occurs —assesses student mastery of objectives —criterion-referenced —data on outcomes
Revision of Instruction and Materials	—based on guesswork or availability of new material —occurs intermittently	—based on evaluation data —routinely occurs

Requirements for Using ISD

Development costs. Now, what of the requirements and limitations associated with instructional systems development? As you may expect there is a cost associated with the usual benefits from ISD. Because of the effort that must be placed in the analysis phase of ISD and in the initial development of training materials, the upfront costs of ISD are greater than the upfront costs of traditional instruction. Thus to engage in an ISD effort will require additional monies *early* in the training cycle to offset the development costs. Later the per trainee costs are usually lower than in traditional instructional methods. This shifting of expenditure to the early phases of a project is typical in capital-intensive projects such as in the introduction of technology. In traditional labor-intensive projects the expenditures are more level over time. The inability of most schools to look at some expenditures as investments accounts in good measure for the lack of ISD at the public school level. Business organizations, naturally, are quite familiar on the other hand with the concept of investing monies now in order to save expenses later, thereby creating large returns on the investment. Unless upfront costs are seen as investment funds, it is unlikely that ISD will be authorized.

Climate. If ISD projects are to be successful, then a supportive management climate is essential. If the management is more concerned with *results* from training as opposed to participation in training programs, then ISD projects should flourish. Management must also understand the time and costs required for the upfront analysis. Individuals who are new to instructional systems development may find it somewhat abstract because it takes a long time before they see an actual training session. If management is pleased and satisfied just because employees went to a training program, regardless of what they learned, then some of the incentives for ISD are absent. It is when management holds training accountable for improvements in employee performance that ISD projects will be more successful. An orientation towards outcomes expected from training is important in ISD. If the time employees spend in training is considered valuable, then ISD will receive more support because the typical finding is that ISD reduces training time. When time and performance are considered important, then instructional systems development efforts are more likely to succeed.

Emphasis on outcomes. When training programs are viewed as means to an end rather than the end itself, instructional systems development will be more effective. If training is perceived as good in and of itself and that employees should participate in training programs with no other particular purpose in mind, then the benefits associated with ISD have limited value. In turn, a poorly conceived and conducted training program could be judged successful if we considered training to be an end in itself. There would be no incentives for developing a better training program. Likewise, if a policy called for each employee *to complete* a five-hour course on job safety rather than *to demonstrate* that afterwards they knew appropriate job safety procedures, then there seems little reason to justify ISD efforts. ISD is more likely to succeed when the training program is held accountable for producing actual results in terms of on-the-job performance, not just for providing courses. In turn, these performances are needed to support the organization's mission. The ability of carefully designed training to meet an organization's goals is the key.

In addition to this (1) supportive management climate and (2) front loading of funding, instructional systems development efforts require that (3) the goals or objectives for the training programs be able to be specified in advance. Briggs (1982) indicated that ISD was more appropriate when:

- objectives could be identified in advance
- predesigned materials were to be used
- empirical data on effectiveness were desired
- the essential objectives were skills or information

He viewed ISD as equally appropriate for group or individualized instruction but cautioned that the instructors must be willing to follow prescribed plans.

Resistance to ISD. It is not unusual to experience resistance to instructional systems development, especially from some of the instructional staff who have become accustomed to determining the instructional content and presenting this content, usually by the stand-up lecture. Certain people sometimes perceive threats from using ISD; they see a diminution of their role as sole possessors of knowledge. They also have a hard time believing that people who have not done the job themselves can understand the needs adequately to make training and content decisions. If the state of knowing about some phenomena is to be protected and reserved for the privileged few, then ISD should be a threat, since it attempts to transfer knowledge widely. If we have special status as a result of certain knowledge we possess, then any attempt to make this knowledge more widely available would likely reduce our status. Ironically, some teachers might not want to teach so well; they may want to maintain a gap between themselves and their students to preserve their special status or maintain power. Obviously, such feelings inhibit instructional systems development. It is appropriate to expect some resistance in implementing ISD, as most training departments are run by functional people (practitioners) rather than by professionally trained educators. When a training department does have educators, it is likely they were hired by top management but must work daily with mid-level people who may not understand their role. To diffuse resistance it is important to conceptualize the work as a "team effort," as a partnership. Both the educator and the content expert have roles to play, each in support of the other. The educator offers a systematic approach to training development; the practitioner brings knowledge of the content.

Staff development. If an organization is to use instructional systems development effectively then persons trained in the ISD process are essential. Should an organization adopt an ISD approach to their training, then an early step is the hiring of persons with knowledge and expertise in the use of ISD, or the training of the current staff in ISD. Successful ISD projects require a skilled staff that can adequately conduct the many tasks or steps required in an ISD process.

When to Use Instructional Systems Development

There has been much discussion about whether the usefulness of ISD should be limited to situations that involve the teaching of simple information or motor skills rather than the teaching of problem solving, attitudes, or other, more complex instructional outcomes. It is easier, of course, to deal with the teaching of an isolated piece of information or a simple motor skill. We doubt, however, that ISD is limited to such outcomes.

The difficulty in using ISD in situations that largely involve the teaching of attitudes or "softer" skills is the requirement in ISD for prespecification of outcomes. It is difficult to be precise about many of the fuzzier outcomes we say we are teaching, things like appreciation of cultural differences, tolerance for co-workers, or a "good attitude." Perhaps the instruction on these topics is like the temperance lectures of days gone by. We can talk and talk about such matters and require our students or trainees to attend but rarely do we know what may be accomplished. In fact, we may be a bit scared to look. We are getting better, though, at working toward these affective objectives, and in this regard the work of Martin and Briggs (1986) is invaluable.

But the debate continues between ISD oriented people who want to define such outcomes and traditionally oriented people who insist that what they are teaching defies definition but that people should simply attend because it is "good for them." Traditionalists argue against attempts to define instructional outcomes by saying that this is a reductionist approach that misses the essence of what they intend. It is easier to define and measure what is meant by multiplication of two digit numbers than what is meant by respect for our heritage. However, if teaching respect for our heritage is important, then we should expend effort to describe what we mean and how we can come to know if we have taught it. ISD seems equally appropriate in the affective area, if just a bit more difficult to carry out.

This chapter has tried to provide a realistic perspective on the benefits, requirements, and limitations of instructional systems development. The theme of this book is that ISD can be used effectively in developing training programs. There are some limitations and requirements associated with ISD; it is not realistic to expect it to be equally successful in all situations. It is difficult, however, to deny the effectiveness of the ISD process.

Under the proper conditions, investment in ISD is indeed a wise use of organizational funds. It is an investment in the future capabilities of the organization.

Chapter 3

The ISD Model

The following is a generic model that describes the tasks and steps that are generally performed in designing training programs in large organizations. There are many variations of the systems approach. However, all ISD models basically have several things in common. First, it is process oriented and not product oriented. All too often, training programs are "sold" as a given program or course without a thorough analysis of the problem to determine if the "product to be sold" is in fact the correct kind of training. Second, this approach is systematic and deliberately planned. Third, there are five stages that are similarly entitled to describe similar purposes. Fourth, output from one task and phase generally acts as input for the next. Finally, this process has built-in controls for review and required revision.

The systems approach generally begins with a FRONT-END ANALYSIS in which training needs are analyzed and identified for different groups of trainees within the organization. This phase is followed by the DESIGN phase, in which goals and objectives are developed and used to determine appropriate content as well as what methods and media to use to deliver the training. Testing procedures are established in the design phase and are also based on the learning objectives. The third phase is the actual DEVELOPMENT of the training materials, which includes content that the trainer will present as well as handouts and exercises for participants and audiovisual materials that will be used to support the trainer's presentation. The IMPLEMENTATION phase follows and is the phase where the training is conducted. It generally begins with a pilot where materials and methods are formatively evaluated before they are made ready for mass use. Instructor training may also take place at this point. The final EVALUATION phase assesses the training at the end of each presentation to determine trainee mastery of the instruction as well as the appropriateness of the training itself. A second type of evaluation generally occurs some months after the training and is designed to follow-up the trainees' use of what they learned in order to determine overall program success.

Our purpose is not to offer a "cookbook" of how and what to do in designing training programs. Instead we have attempted to describe the process by stating the tasks that are generally performed in a sequence as they generally occur. The model is presented as a guide to help you understand the logical flow of events. These tasks are temporally ordered, as the output from one task tends to be used as input for the next. In turn, we have stated tasks as action statements to make them a bit clearer and more concise. We have additionally tried to include the kind of steps required to design large scale training programs that involve a series of courses for several different groups of trainees which may require several years to develop and implement.

In addition to following a systems approach, it is important to note that most training programs are designed by teams of training professionals. These teams may include the training designer who provides the educational expertise, a content specialist who knows the functional background required for a given training topic, and a manager who represents the vested interests of the organization and knows what resources and constraints must be factored into the decision-making process. In some organizations the manager may have a background in education or human resources and in others he or she may be a functional expert. Some tasks are performed by one person whereas others may be performed by several. Team members may additionally be different in different phases of the project. Those individuals who are initially involved in the Front-End Analysis portion of the program may not, for example, be the same ones responsible for completing tasks in the Development phase.

We will now attempt to walk you through the tasks in our model. Keep in mind that most designers will customize this ap-

The ISD Model

proach according to their unique situation and needs. Certain tasks within a given phase, for example. may be completed at the same time or performed in less detail than described in this text. Other tasks may be considered unnecessary in a given situation and not performed at all. It is also important to note that the process is not one hundred percent linear. Some designers may do several tasks at once or not in the sequence described in the following outline. Succeeding chapters will help you determine how to best use the model. In turn, they will focus less on the step-by-step aspects of the process, and more on the concepts and issues that underlie the procedures themselves.

At this point, we will begin by listing the tasks for each of the five phases followed by a general discussion of their functions.

ISD Model

I. **Front-End Analysis**
1. Respond to request for training assistance.
   - 1.a. Meet with client to gather initial information on history and scope of problem.
   - 1.b. Explain scope of your services and methodology.
   - 1.c. Gather initial information about the organization's mission and environment.
2. Negotiate assessment plan.
   - 2.a. Develop plan for sources, instruments, methodology, time limits, field procedures, expected balance of quantitative and qualitative data, and criteria for decision-making.
   - 2.b. Negotiate assessment plan and gain management commitment.
   - 2.c. Document trade-offs and risk of invalid findings if negotiated plan differs greatly from the ideal.
3. Collect data on overall problem.
   - 3.a. Select and or develop data collection instruments.
   - 3.b. Gather information.
   - 3.c. Document collected data by preparing charts, tables, etc.

4. Analyze incidence of problem.
    4.a. Calculate quantitative and qualitative data.
    4.b. Compare data against preferred norms to determine performance gaps.
    4.c. Analyze importance of performance gaps.
5. Determine probable cause(s) of performance gaps.
    5.a. Distinguish between needs that can be solved by training and those related needs that must be addressed by a change in organizational procedures or policies.
    5.b. Document and discuss training-related needs that must be addressed by the organization.
6. Prioritize identified training needs.
    6.a. Link training needs to existing or new jobs.
    6.b. Identify job components by conducting job task analysis for each specified job.
    6.c. Assess capability of current job incumbents to complete tasks.
    6.d. Prioritize criticality of tasks that require training.

II. Design
MACRO DESIGN—Program and Curriculum Levels.
1. Determine program level goals and objectives.
    1.a. Link educational component to support of organization's mission.
2. Determine curriculum goals and objectives.
    2.a. Convert training needs into training topics.
    2.b. Group training topics into curricula by subject matter or by personnel group.
    2.c. Identify courses to be developed for each curriculum.
    2.d. Prioritize and determine order of course development.
3. Conduct initial audience analysis.
    3.a. Collect data on audience background, previous training experiences, availability to attend training, etc.
4. Determine feasible options for methods and media.
    4.a. Analyze requirements for training topics and target audience(s).
    4.b. Consider resources and constraints.
    4.c. Prepare list of feasible options.

MICRO DESIGN—Course and Lesson Levels
  5. Specify course goals.
     5.a. Identify terminal behavior for each training topic.
     5.b. Conduct goal analysis to ensure clear goal statements.
  6. Specify learning objectives.
     6.a. Conduct instructional analysis to identify prerequisite skills.
     6.b. Determine what prerequisite skills target audience has yet to learn in order to reach the terminal behavior.
     6.c. Convert remaining prerequisite skills into learning objectives.
  7. Determine appropriate content to operationalize learning objectives.
     7.a. Write high level outline.
  8. Develop test items for each learning objective.
     8.a. Write test items.
     8.b. Validate test items to resolve validity and reliability issues.
  9. Select methods and media required to operationalize each lesson.
  10. Conduct make/buy decision.
     10.a. Determine if possible to purchase existing courses, lessons, and audio-visual materials from vendors.
     10.b. Plan for lesson development if necessary.

III. Development of Content
  1. Complete specifications for each lesson.
     1.a. Write detailed outline.
     1.b. Develop conceptual sketches for audio-visual aids.
     1.c. Validate with subject matter expert if developed by training designer.
  2. Develop lessons and activities.
     2.a. Write lessons and activities.
     2.b. Review/edit copy if written by subject matter experts.
  3. Conduct formative evaluation—The Walkthrough.
     3.a. Walkthrough materials with additional subject matter expert(s), management and one or two members from target audience.

3.b. Revise as needed.
3.c. Produce materials for pilot.

IV. **Implementation**
1. Develop overall implementation plan.
    1.a. Agree/confirm plans with management on schedule, logistical and instructional requirements, and selection of success criteria for pilot and subsequent stages for the program.
2. Prepare instructors
    2.a. Develop instructor training if needed.
    2.b. Train instructors.
3. Conduct formative evaluation—The Pilot Test.
    3.a. Arrange for site and participants.
    3.b. Conduct and evaluate pilot.
4. Revise and produce training for mass use.

V. **Evaluation**
1. Collect summative data at end of each course.
    1.a. Determine participants' ability as individuals and as a group to master learning objectives.
2. Analyze summative data for participant, instructor, and course success by individual presentation and across presentations.
3. Revise course as necessary.
4. Collect follow-up data to determine program success.
    4.a. Negotiate assessment plan and schedule with management.
    4.b. Develop and validate needed instruments.
    4.c. Collect data.
    4.d. Analyze program success over time.
5. Revise program as needed.

Discussion of the Model

Front-End Analysis

The purpose of this phase is to determine how training can benefit society, an organization, or a department within an organization. This phase is thus one of the most complex and can require a great deal of time to complete. It begins with little or no assumptions about what training is required. Rather, it attempts to first analyze the "big picture" by linking perceived problems to performance deficiencies. Such performance problems may presently exist because individuals do not have the needed knowledge and skills to carry out present or planned tasks required to reach certain societal goals, mission statements, or business plans. In some instances training may not be the answer. Instead, findings may suggest changes in procedures or policies. Once training needs are identified for various groups, they are prioritized in accordance with a predetermined set of criteria.

The first step is to generally meet with the client to gather information on the history and scope of the problem. Most training programs are designed for a specific organization. This phenomenon is true even when designing a very large scale program designed to address societal problems such as environmental control and protection. If the designer is already a member of an organization's training department, then the "client" may be a supervisor from another department or a highly placed official within the organization.

Requests for training may be highly specific, such as "we want a course in stress management" or fairly vague, such as "productivity has been low lately, and we think that our people need some kind of training to improve their efforts." Training requests may not only range from vague to specific, but the nature of the request itself may also vary. In some instances the designer may be asked to design one or more courses that must be compatible with an existing training curriculum that is part of a larger program that functions as a part of the strategic plans for that department or organization. In other cases, the designer may be asked to start at the beginning and create plans for a training program that can support the organization's strategic efforts to carry out its mission or business plan.

As needed, the designer should explain the scope of his or her services and training methodology. Information should also be gathered about the organization's or department's mission and environment. This information is used to develop and negotiate a needs assessment plan. It is important to gain client approval and commitment to facilitate the research and ensure the appropriateness of the plan. Negotiations should include agreements on sources, instruments, methodology, deadlines, field procedures, and the expected balance between quantitative and qualitative data. In addition, criteria as to what constitutes a training need and how identified needs should be prioritized should also be pre-determined. If the client disagrees with the recommended methodology or is reluctant to spend the money, time, or make needed personnel available, then trade-offs and risks of conducting an assessment that yields incomplete or untrue data should be documented before finalizing the agreement.

The next step is to collect data on the overall problem. It is generally best to gather data from several different sources and to use a variety of methods. Suggestions for data collection are covered in Chapter 4. Once data collection instruments are selected or developed they should, of course, be pretested to ensure suitability, and collected data should be documented by preparing charts, tables, graphs, etc.

Data are next calculated to analyze the incidence of the problem. In this part of the assessment process, performance problems are first evaluated by comparing them to ideal expectations. Performance expectations may be based externally on industry norms or government legislation or internally on standards set by top officials in the department or organization. Second, the frequency and criticality of identified performance gaps are identified to determine just how serious they are.

To determine the probable cause(s) of these performance gaps, the designer must first distinguish between needs that can be solved by training and those related needs that must be addressed by a change in organizational procedures or policies. Questions and procedures for analyzing this kind of analysis will be covered in Chapter 5. Performance problems that cannot be solved by training should be documented and discussed with the client. In some cases, failure to make the necessary organizational changes will hinder training efforts.

Finally, identified training needs are prioritized. In many cases, there may not be enough resources or time to develop training for all identified needs. Additionally, the organization may not consider certain performance problems serious enough to merit the cost of developing training to close identified gaps between the current and ideal performance. In order to make this decision, training needs are first linked to existing or new jobs. Next, the components of each job are specified by conducting a job task analysis. The capability of current or potential job holders to complete identified components is then assessed to determine where and if training is required. Training topics are then identified and prioritized according to their importance and effect on the organization's overall problem.

Design

In the design phase, goals and objectives are developed which are then used to select content, instructional strategies, and testing procedures for each training topic identified in the Front-End Analysis. The design process can be thought of as occurring on four different levels: (1) Program, (2) Curriculum, (3) Course, and (4) Lesson. The program and curriculum levels are associated with a MACRO type of organizational analysis and decision-making. At this point training is linked with the strategic plans of the organization, and a series of course needs are identified for different groups of trainees. Levels three and four comprise a more in-depth or MICRO type of planning. At this point, decisions are based on instructional theory and research. Thus they are concerned with the learner's ability to understand and remember material presented in the individual course and lesson.

MACRO DESIGN. Program level goals are first developed to describe how a training program can contribute to the achievement of societal goals, mission statements, and business plans. Their objectives outline tasks and personnel required to support strategic planning efforts. Information from the task analysis portion of the front-end analysis phase is then used to determine the present capability of individuals to carry out their roles. In turn, areas where training is needed are specified and organized into curricula. Curricula may be organized by subject matter as in "management development courses" or by personal groups as in "career development planning." Once training needs can be placed into curricula, they are further organized into individual courses which are then prioritized and scheduled for development and implementation. An additional component of macro design is the initial selection of methods and media that can be used to deliver the training. This type of decision-making is called a feasibility analysis and is driven by the nature of the training topic, macro-level goals and objectives, an initial assessment of audience characteristics gathered during the front-end analysis, and a consideration of existing resources and constraints.

MICRO DESIGN. Micro design begins with a specification of course goals by identifying the terminal behavior for each training topic that will be included in a given course. The terminal behavior states what the learner is expected to be able to do upon completion of the instruction. Goal statements are written as observable behaviors and a procedure called "goal analysis" can be applied to ensure their clarity. An instructional analysis of the goal is next conducted to identify prerequisite skills required to reach the terminal behavior. A second, more in-depth audience analysis is conducted to determine what prerequisite skills the target audience has yet to learn in order to reach the terminal behavior. Remaining prerequisites are converted into learning objectives and used to develop a content outline for individual lessons and finalize the selection of methods and media. Designers next determine if it is possible to purchase existing courses, lessons, and audio-visual materials from vendors that will meet their requirements, or if training materials need to be developed that can be used to pretest learners, ascertain individual mastery once the instruction is completed, and assess the appropriateness of the course design.

Development of Content

In this phase the training materials are developed. Designers often work closely with functional specialists who review and may even write the first draft of the materials. Designers may also seek the artistic and production expertise of audio-visual specialists if they are designing media in formats with which they are unfamiliar.

The ISD Model

Specifications developed in the design phase are first completed for each lesson. Tasks include writing a detailed outline and the development of conceptual sketches for audio-visual aids. The outline and sketches are generally reviewed by other members of the design team before the actual writing begins of instructor scripts, handouts, and participant exercises. Generally drafts are again reviewed and revised before a formal "walkthrough" takes place. Suggestions for working with members of a design team are covered in Chapter 11, and a discussion of formative evaluation is included in Chapter 13.

A Walkthrough is part of the formative evaluation process designed to identify problems and correct them before the materials are produced for mass use. The second type of formative evaluation occurs when the training is piloted in the first stage of the implementation phase. In a typical walkthrough, materials and audio-visual sketches are distributed to the client management, functional experts, and one or two representative members from the target audience. After their initial review, the designer meets with members of the review team and walks them through the materials and discusses their suggestions for revision. Materials are then revised and produced for the pilot. It is important to note that a walkthrough is different from a pilot where the training is actually conducted as if it were a typical situation.

Implementation

It is in the implementation phase that the training is actually conducted. Most large-scale programs are implemented in stages. They generally begin with a pilot followed by an expanded effort designed to include larger numbers of trainees as well as those members of the target population that are likely to be more resistent to change. Staged approaches are useful because they permit a chance to formatively evaluate the program and thus present less risk to decision makers. They also promote a readiness for change through demonstrable results and encourage clients to gradually become more involved in the implementation of their own program. Strategies to promote the kind of planned change that results from training efforts are discussed further in Chapter 12.

The implementation of any training program first requires an agreement with client management that covers the schedule and logistical and instructional requirements needed to conduct the training. Often, tentative plans are developed at the end of the front-end analysis phase once curriculum requirements are identified and a schedule for course development can be determined. Before beginning the actual implementation process these plans are again reviewed for confirmation and revised as required. As part of the plan, criteria should be pre-determined to evaluate the success of the pilot before continuing on with the program. Likewise, decisions as to what constitutes readiness for the implementation of successive stages should also be determined beforehand.

Instructor training may also take place during this phase. In many instances the instructors are not a part of the design team and may thus require some assistance in the areas of instructor skills as well as in acquiring familiarity with materials and the intent of the training. Such presentations are usually designed and written during the development phase and presented as required throughout the implementation of the program.

Evaluation

Two types of evaluation actually take place in this phase. The first is called summative evaluation, as it is conducted at the end of each training presentation and thus acts as a kind of summary of that course session. Summative data collection includes the assessment of learner mastery of the instruction as well as the appropriateness of the training design. The second type of evaluation occurs some time after the instruction and is called follow-up evaluation. Its purpose is to evaluate how and if the training is being used by the participants and is used to determine the overall success of the training program.

The course is first evaluated at the end of each presentation to assess the ability of individual participants to master the training. Testing procedures are based on the learning objectives and permit the identification of those parts of the training that were difficult for the individual to learn. Learning difficulties can then be addressed by prescribing additional instruction pertinent to certain objectives. The data are also assessed as a group to determine the appropriateness of the design. If large numbers of the group had difficulty achieving certain objectives, then there may be something wrong with the instruction itself. The instructor's effectiveness is also considered at this point. Generally, participants complete evaluation forms to ascertain their overall response to the instruction and the instructor. Summative data are additionally reviewed across presentations to see if similar patterns are present that indicate a need for course revisions.

Follow-up data may be collected some months as well as years after participants attend a given training session. While many evaluators may be associated with the same organization as the training designers, they are seldom the same individuals who actively worked on the design of the program. This kind of distance permits a more objective assessment of the findings. Evaluators should first negotiate an assessment plan and schedule with the client followed by the development and validation of collection instruments. Once the data are gathered, they are then analyzed over time to determine overall success and the program is revised as needed.

Summary

This chapter has attempted to describe the systems approach to the design of training as employed in large organizations. We have presented a five-phase model that can be used as a guide to understand the progressive flow of events and decisions. Tasks were presented as action statements and temporarily ordered to clarify how output from one phase can be used as input for the rest. We have additionally tried to explain that training programs generally constitute a team effort and have indicated places where different kinds of expertise are especially valuable.

Our purpose has thus been to present a model that should be used and adjusted according to the training designer's unique environment and needs. Not all designers will use it in the same way. Some may perform certain tasks simultaneously, or in less detail that what is described in this text or perhaps not at all. Succeeding chapters will further enable the training designer to decide how to use the model by describing the rationale and issues that support the ISD process.

Chapter 4

Needs Assessment

The need for accurate and accountable training programs increases as the demand for training grows and becomes a strategic part of an organization's plans. Many organizations view making the right decisions about how to train what trainees on what knowledge, skills and attitudes essential to insure a return on investment in training. A thorough needs assessment that analyzes and traces the source of identified needs is required to help reduce the risk of funding inappropriate programs. This is often called a front-end analysis and should come at the beginning, prior to developing and conducting the training.

The Discrepancy Model

Needs assessments are the systematic identification, evaluation, and prioritization of identified needs. A basic discrepancy model is generally applied to identify and assess the nature and extent of these needs. Using this approach, the ability of individuals or groups to meet performance expectations for a given job or task is first analyzed. If job holders are unable or unwilling to meet performance expectations, the differences between their current capabilities are measured and translated into performance gaps to form the basis for determining additionally required knowledge, skills, and motivation.

Kaufman (1976), Witkin (1977), Kaufman and English (1979), Fessler (1980), Lane, Crofton, and Hall (1983), and Jones and Clay (1984) all define the needs assessment process as a method for identifying gaps between present and desired outcomes. Training needs are based on the discrepancies between the expected performances and the actual, or current, performances.

> PERFORMANCE EXPECTATIONS
> −CURRENT PERFORMANCE CAPABILITY
> _____
> =TRAINING NEEDS

Kaufman (1976) describes needs assessment as a three-step process:
1. Determine all relevant gaps.
2. Rank identified gaps in priority order.
3. Select the highest priority gaps for closure.

Needs Assessments are generally conducted on different levels and may begin by analyzing the target audience's societal and organizational environment. Since training programs are designed to enable persons to function effectively in an occupation, on a specific job, or as a member of society, the desired outcomes must be derived by analyzing the environment in which the target audience lives and works. The next level generally begins by determining discrepancies between desired organizational goals and current levels of performance. These discrepancies are then prioritized and form the basis for training goals. In the third level a job analysis is used to identify the specific knowledge and skills necessary for successful job performance. Data gathering techniques are varied and must be tailored to specific situations. They may range from telephone interviews to performance tests and may include a review of management requests, expert statements, and soliciting information from the intended trainees as well as individuals affected by their performance, such as supervisors and clients. Once the data are gathered, they are analyzed to determine what needs can be directly addressed by training. All too often, training is viewed as a panacea for all ills. A performance analysis is conducted to determine if needs represent a lack of knowledge/skills and inappropriate attitudes or if needed change in organizational policy and procedures is required. Finally, information on organizational resources and constraints are considered in making choices on how to design and deliver the desired training.

Needs Are Multi-Dimensional

As we have seen, Kaufman's discrepancy model characterizes needs as the gap between observed results and required results or the difference between what is *vs.* what should be. This simple calculation provides a clear base for an evaluator to begin his or her assessment. However, once on site, the waters are easily muddied as the evaluator often finds that the concept of need is influenced by additional factors. Additional theorists suggest that once a basic discrepancy calculation is made it must be reviewed in light of the organization's environment. Coffing and Hutchinson (1974) advise us to pay attention to the influence of gatekeepers and decision-makers. They describe the concept of need as who needs what as defined by whom. It is not uncommon to receive a request from an official to develop a training program for a predetermined group in a predetermined content area. Such requests can restrict or even block a thorough needs assessment. Gaining the support of those in charge is essential to the success of any analysis. The concept of who needs what is echoed by Stufflebeam (1977) who cautions that different types of need exist for different levels of individuals or groups. The ability to respond to those needs is, in turn, influenced by available resources and constraints. In collecting information on needs, it is also important to remember that needs are not static. Scriven (1978) points out that a difference exists between short-term, long-term, and maintenance needs. These differences are especially noteworthy in working with a growing organization or one that is subject to changing laws and regulations and changing trends within the industry. Finally, Scriven and Roth (1978) believe that needs must not be confused with wants. Identified gaps must reflect an unsatisfactory state of affairs. An identified need should indicate that an individual, group, or system, may be harmed or indisposed if the need is ignored. For example, the ability to speed-read may not be essential for salesmen to meet their quotas within their allotted deadlines. On the other hand, training in time management may permit salesmen more time to make new client calls and therefore be more capable of meeting their quotas.

Needs Originate from Different Places

Training needs can generally be traced to multiple sources which are both internal as well as external to an organization. Needs may be referenced to performance expectations set for a given task, an organization, or society as a whole. Problems are often first acknowledged through failure to meet a critical event. The notion of a critical event is associated with failure to meet a given set of performance criteria. Note the following examples:
- Our people cannot perform well-enough on task X.
- Our profits are lower than last year.
- The incidence of infant mortality is too high.

It is important to realize that needs associated with any critical event do not exist in isolation. Any set of needs is related to other levels of need which can be both inductively as well as deductively traced. The following set of questions can be used to analyze the path and differing need levels. This paradigm begins with the big picture and narrows its focus with each successive set of questions.

1. COMMUNITY AND SOCIETY
How can an educational program improve general living conditions?

Example Conditions:
    Standard of living.
    Life expectancy.
    Disease control.
    Healthy GNP.
    Unemployment.
    Peace.
    Educational standards and access.

2. ORGANIZATION
Is the targeted organization capable of responding to a societal problem?

Example Capabilities:
- General belief in the need to respond to the problem.
- General policy regarding the problem.
- Technical and managerial expertise.
- Manpower and money.
- Open communication channels throughout the organization.
- Common perspective on the problem.
- Ability to command and share resources with other organizations.

3. TASK

Are current job holders able to perform tasks required to carry out the organization's mission?

Example Problems:
- Unreasonable performance expectations.
- Lack of needed knowledge, skills, and motivation to do the job.
- Lack of time to participate in training activities.
- Inadequate reward system and other organizational constraints.

The following example illustrates how these questions can be used to understand the source of training needs.

Problem One

SOURCES
COMMUNITY/SOCIETY:
 High incidence of heart disease

ORGANIZATION:
 Local public health department will design and deliver the training program.

TASKS:
 Design Course Materials/Raise needed funds to implement program on on-going basis.

CRITICAL EVENTS
COMMUNITY/SOCIETY:
   Average number of heart attacks is too high.

ORGANIZATION:
   Difficulty in gaining funding for program. No experience in the design of training programs.

TASKS:
   Assigned personnel are anxious about designing programs. They know the content but are unfamiliar with how to design training. Fund-raisers exhibit poor attitudes due to poor track record in gaining money.

NEEDS
COMMUNITY/SOCIETY:
   Community members need to practice better dietary habits to reduce risk of heart disease.

ORGANIZATION:
   More money for program expertise in the design of training programs.

TASKS:
   Personnel who know needed content as well as how to design training programs. Personnel who are more confident and capable in raising needed funds.

Problem Two

SOURCES
COMMUNITY/SOCIETY:
   Decreased standard of living for corporation X employees.

ORGANIZATION:
   Department Y within corporation X.

TASKS:
Perform tasks needed to meet quotas on time supervision of employee performance in order to ensure higher productivity and less turnover.

CRITICAL EVENTS
COMMUNITY/SOCIETY:
Employee layoffs and fewer raises—steadily declining company profits.

ORGANIZATION:
Unable to meet quotas on time; high personnel turnover.

TASKS:
Many managers appear uncertain and ineffectual. Most have little or no managerial training and possess uneven knowledge of the tasks they are to supervise. Few managers have time to gain needed training. Many employees express frustration due to poor supervision and lack of guidance in completing tasks.

NEEDS
COMMUNITY/SOCIETY:
Increased corporation profits are needed to ensure more raises and fewer employee layoffs.

ORGANIZATION:
Higher productivity in a more timely manner—more stable work force.

TASKS:
Managers who exhibit good management skills as well as performance level knowledge of the tasks they supervise. Better supervised employees.

In this example the problem did not lie with the employees responsible for performing those tasks most closely connected with the problem. The real problem originated with their bosses, who did not know how to manage very well. This problem also illustrates how identified needs often require multiple solutions. For example, the need to increase competency in new managers may not only require knowledge and skills training in managerial and task/level content. Time away from the job may be required additionally to attend training workshops. Peer tutoring by a more senior manager also may be needed. This second set of solutions cannot be directly addressed by training. While the issue of more time for training will influence effectiveness, it must be met by a change in the organization's procedures or policies. This example is illustrated in Figure 4.1.

Data Are Used for Different Purposes

Information from a needs assessment provides the foundation on which the design of the training program rests. As with most architectural designs, if the foundation is unsteady, the remaining structure will be weak and likely to fall. Thus input from a weak analysis will lead to poorly designed goals and objectives, which will lead to a poor choice of instructional content, methods, and media, etc. In a general sense, data are used to support the entire training cycle. Data that yield information about the organizational environment, the current state of performance capabilities, the target audience(s), and training policy are needed to ensure a strong design. Most specifically, needs assessment data can be used to develop training plans for either an entire curriculum or an individual course.

General Information: The following general information is needed to design training:

Organizational Environment
PURPOSE: To describe the environment in which the problem exists.
SAMPLE DATA: What is the organization's mission, business plans, and products? How does the failure to meet certain critical events affect the organization's ability to meet its goals? Is the organization currently capable of meeting its goals? Who are the gatekeepers, decision-makers, and stake-holders that will influence decisions about training? Who perceives a need for training within

Needs Assessment 47

Figure 4.1 Multiple Solutions for an Identified Need.

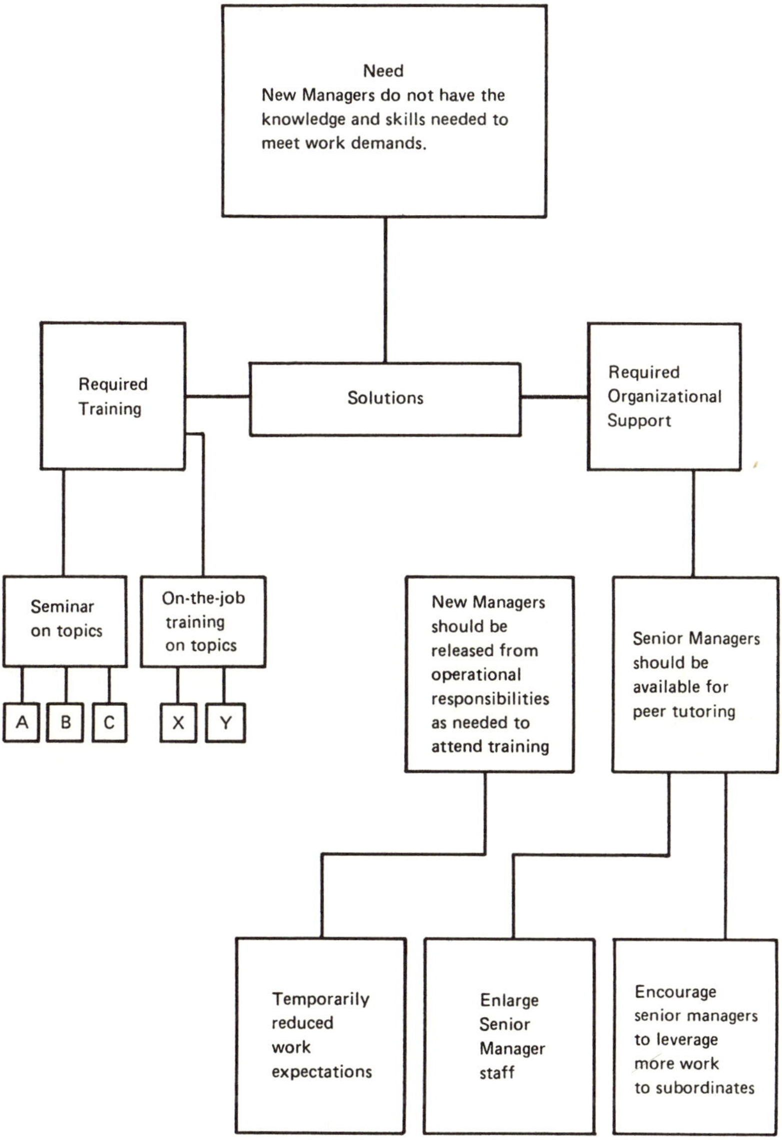

```
                    ┌─────────────────────┐
                    │        Need         │
                    │ New Managers do not │
                    │ have the knowledge  │
                    │ and skills needed to│
                    │ meet work demands.  │
                    └──────────┬──────────┘
                               │
          ┌────────────────────┼────────────────────┐
          │                    │                    │
    ┌─────┴─────┐        ┌─────┴─────┐        ┌─────┴─────┐
    │ Required  │────────│ Solutions │────────│ Required  │
    │ Training  │        │           │        │Organizational
    │           │        │           │        │  Support  │
    └─────┬─────┘        └───────────┘        └─────┬─────┘
          │                                          │
     ┌────┴────┐                                     │
     │         │                                     │
┌────┴───┐ ┌───┴────┐  ┌──────────────┐      ┌──────────────┐
│Seminar │ │On-the- │  │New Managers  │      │Senior Managers│
│on      │ │job     │  │should be     │      │should be     │
│topics  │ │training│  │released from │      │available for │
│        │ │on topics│ │operational   │      │peer tutoring │
└─┬─┬─┬──┘ └──┬──┬──┘  │responsibilities      └──────────────┘
  A B C       X  Y     │as needed to  │
                       │attend training│
                       └──────────────┘
```

| Temporarily reduced work expectations | Enlarge Senior Manager staff | Encourage senior managers to leverage more work to subordinates |

the organization? Who has what kind of expectations for training? What is the managerial network? Will decision-making practices and the manner of assigning tasks impact on training? What organizational constraints will limit training's effectiveness?

IMPACT ON TRAINING CYCLE: Helps to prioritize identified needs. Helps determine performance expectations for prioritized needs. Helps uncover non-training needs that may influence training effectiveness. Helps identify potential difficulties in gathering required needs assessment data.

Baseline Performance Data (Job-Task Analysis)

PURPOSE: To establish a base for measuring change over time. To determine discrepancies between ideal and current performance.

SAMPLE DATA: What is the present level of performance capability? What is the difference between present performance and the organization's expectations for the job? What is the socio-economic status of the job? What are the political issues associated with the job? What is the life cycle estimated for the job? Will changes in the job require future adjustments to the training? What knowledge and skills are needed to be able to complete the job? Does this information need to be memorized or can it be looked up and referenced in the course of completing the job? Can job procedures be improved? For example, can steps be simplified, eliminated or resequenced to produce increased output, reduced error, easier work, etc.? What personnel groups impact the performance of the task? For example, who performs the task and who manages it? What are the departmental and co-worker interfaces?

IMPACT ON TRAINING CYCLE: Translate performance gaps into types of training needed. Establish training goals. Identify groups/individuals to be trained.

Description of Target Audience(s)

PURPOSE: To discover who the target audience is. To assess the importance of group and individual differences.

SAMPLE DATA: What skills and knowledge do they bring to the job? How homogeneous or heterogeneous is the target group(s)?

Should they be trained in groups or trained individually? What past experiences have they had with training? What is the audience's familiarity with certain training methods and media?

IMPACT ON TRAINING CYCLE: Determine what groups/individuals can be trained together. Establish observable, measurable, behavioral objectives for a given group or individual. Establish performance criteria for criterion-referenced tests. Input for methods and media selection.

Training Policy

PURPOSE: To determine training procedures and policies.

SAMPLE DATA: What is the expected life cycle of the training program? How static is the content? Will the program be available to new and newly promoted personnel on an ongoing basis or is this a one time session? Should the administration of training be centralized or decentralized? How important is this type of training to the organization? Should certain personnel groups be trained first to help create a positive atmosphere for program dissemination? Who will be responsible for administering and revising the program as needed? Who will deliver the training? Do they know how to train others and are they thoroughly familiar with the content and expectations for the program? What constraints need to be considered in terms of time and cost? What material and personnel resources will be available to help in the design of training? For example, will internal content expertise be available, or will outside sources need to be consulted? What kind of training presently exists? Can any existing training be revised or updated to meet current needs?

IMPACT ON TRAINING CYCLE: Determine procedures to use in developing and delivering training. Determine need for outside resources. Determine procedures for disseminating training.

SPECIFIC PURPOSES: Data are generally used to first develop curriculum plans. Curriculum planning involves the systematic identification of training needs of individuals or groups within an organization. This type of an assessment yields an understanding of the big picture and can be used to plan a series of courses that may be developed and delivered over a period of several years.

Curriculum assessments are generally less detailed than those used to design individual courses because some course development will probably be deferred until later years in the program. Also, projected courses may be eliminated if changes in the business plans or mission statements occur. Needs assessment data used to design individual courses add specificity to the earlier identified needs during the curriculum planning phase. The purpose is to determine a set of training needs that are specific enough to allow development of detailed learning objectives for the course. The work effort will depend on the quality, age, and level of detail of earlier data gathering.

What Are the Benefits?

Finally, it is important to describe the benefits that can be gained from conducting a thorough needs assessment:
- Needs assessments can save the cost of developing training that is not needed.
- Clear statements of needs produce clear criteria for assessing learners and evaluating program effectiveness.
- Clearly determined performance needs help identify training solutions that are more accurate and more likely to reduce identified performance gaps.
- When needs are clear the design and development process requires less time.
- Specific descriptions of the target audience lead to more beneficial training since it is targeted to the group that really needs it.
- Needs assessments encourage a commitment to educational planning by organizational decision-makers.
- Training effectiveness can be enhanced by addressing needed changes in organizational procedures and policies.

Summary

As discussed in this section, the information from a training needs assessment is used to determine what type of program to offer for what group of participants, appropriate outcomes, what instructional options (media and methods) to consider and what procedures and policies should guide the program's implementation. A sound front-end analysis can save time and money and greatly contribute to greater accuracy and training accountability. Specific methods for gathering and analyzing data will be discussed in the next section.

Needs Assessment Data Collection and Analysis

Introduction

Needs Assessments should be carefully planned. They require a systematic approach to decrease the risk of invalid findings and to ensure the most effective and efficient use of time, personnel, and money. Needs Assessments are not all alike. Decisions about how to best collect and analyze data are driven by the specific nature of the organization and its problem. For this reason, this chapter will not attempt to describe a step-by-step approach. Instead, questions and issues will be discussed that must be dealt with in developing any Needs Assessment plan.

General Guidelines for Collecting Data

Needs Assessments are not research in the classical sense. The numerous factors that may confound findings make laboratory models difficult to implement. Therefore, more practical guidelines are employed that are not as rigorous in adhering to rules for sampling, reliability, and validity. However, Needs Assessments are like research in that they try to minimize the risk of untrue findings, and so will customize certain principles to enhance the usefulness of data. The following principles are often employed to guide an approach:

- Ensure that problems are examined from a number of perspectives by employing a combination of data sources and collection methods.
- Build in redundancy to increase trust in the data.
- Gather data in a consistent way to reduce the influence of evaluator biases on collected data.
- Allow for both a broad as well as an in-depth look at the problem.

52   *Instructional Systems Development in*

- Solicit organizational input in condu[cting...] ensure collection of relevant informati[on in a] cost-effective manner.

Specific Guidelines for Collecting Data

The following set of questions is a helpful to[ol to] use for planning a data collection approach. There are four sections designed to help answer the following questions:

—What do I already know about the problem (Stage)?
—What data do I need now (Data Needed)?
—How do I find it (Data Source and Collection Method)?
—How much do I need (Sample Size)?

Information required to complete each question is detailed below.

1. STAGE. The process of assessing needs can be organized into three stages:

(a) Key Informant,
(b) Validation, and
(c) Clarification.

Each stage is designed to provide a more in-depth understanding of the problem and its source. Data are summarized at the end of each stage and compared to earlier findings.

Key Informant. This stage presents an opportunity for the training designer to become familiar with key players and data sources. An attempt is made to understand the parameters of the problem and its impact on the organization. In turn possible cause and effect relationships are explored with experts and those closest to the problem. The Key Informant does the following:

—Conducts individual and small group interviews with key officials, experts and personnel closest to the problem.
—Reviews key documentation.
—Initial observation to assess organizational climate.
—Develops rapport; gains needed commitment to do a needs assessment.
—Develops strategy for next collection stage.
—Identify possible difficulties in gathering data.
—Reviews existing training opportunities.

[Sticky note: It appears that the word "stage" is being used in more than one way here — very confusing!]

Needs Assessment 53

Validation. In this stage suspected cause and effect relationships are validated by collecting data from a different, larger, and more representative sample of the target audience. The validation process may first begin by repeating the collection of data in small group meetings (five to seven individuals in a group). Techniques such as the Nominal Group Technique (Delbeq, Van de Ven, and Gustafson, 1975) and the Delphi Method (Linstrone and Turoff, 1975) can be used to reach group consensus to identify, prioritize, and expand on the cause of performance gaps. Structured techniques to reach group consensus are very useful. Not only are they efficient, but also they ensure equal contribution from everyone involved as well as equal weight to all ideas introduced. Cummings and Bramlett (1984) suggest conducting one or more small group meetings with both key informant types (different individuals than those in stage one and target group members who are not widely recognized as specialists in the area of needs that is to be studied). This approach permits validation from recognized experts as well as individuals who are more likely to represent the majority view held by members of the target audience. If significantly different data emerge from what was collected on the Key Informant level, it may be necessary to continue the validation process with a follow-up survey to investigate needs on a wider basis. Surveys can yield the broadest input and can involve one or more of the following methods: questionnaires, interviews, or telephone surveys.

To ensure reliability of findings it is important to structure and ask questions in as similar a way as possible through the use of a structured protocol. It is also important to survey as representative a group as possible. In surveys that include a large number of participants, data analysis will be made easier and more efficient through use of closed ended questions. Closed ended questions are those that call for yes, no, or short answer responses. A more open questioning approach that yields broader and more in-depth answers may be used with smaller groups of respondents. Validation is used to do the following:

—Conduct one or more small group meetings with members of the target audience to secure consensus on the nature of the problem.

—When needed may survey target audience using questionnaires, interviews, tests, and/or other methods.
—Compare summarized data to key informant findings.

Clarification. The final stage clarifies any outstanding questions that may remain once the data are summarized and reviewed from the first two stages. This stage is similar to the Key Informant stage in that information is collected from a small number of sources. Assistance in clarifying data may be had by conducting another round of small group meetings or through interviews with key informants. Clarification is used to do the following:

—Clarify outstanding questions before finalizing data analysis.
—Conduct small group meetings or interviews with key individuals.

2. **DATA.** For any stage in the Needs Assessment process, there are generally four reasons for collecting data. These reasons were described in the previous chapter and are associated with a desire to learn more about the organization and its performance needs, along with the need to determine who the target population is and what training policies will guide the intended program. Below is a listing which gives sample data for each of these four data gathering needs:

- **Assesses Organizational Climate**
    Goals, mission, strategic plans.
    Current operating procedures.
    Current standing in the industry, and/or society; restrictions caused by ecology, politics, religion, or outside interest groups.
    Organization's ability to meet its goals. Where are the gaps?
    Management style.
    Key decision-makers.
    Degree of accountability to outside funding, grants, stockholders.
    Financial commitment to training.
    General morale.
    Employee turnover.
    Criteria for program success.

Is the organization funded by an outside source?
Is this training seen as an isolated incident or is there an ongoing need for training?
Size of organization, sales, number of employees, geographical coverage.

- **Assess Baseline Performance**
    History of performance problem.
    Impact of performance on organizational goals.
    Personnel involved.
    Job descriptions (steps, sequence, etc.).
    Job importance and complexity.
    Morale and turnover.
    Lifecycle of job.
    Organizational constraints on performance.

- **Describe Target Audience**
    Entry-level abilities and skills of the audience.
    Size of group(s).
    Educational background and job history.
    Homogeneous in background/skills.
    Who do they report to? (Immediate superiors, subordinates and clients.)
    Career path expectations.
    Attitudes towards need for better performance.
    Previous experiences with training.
    Learning styles.

- **Determine Training Policies and Procedures**
    The past history of training programs (internal and external).
    Policies and the possibility of revising and customizing existing training procedures.
    Accessibility of trainees (time and locale).
    Deadlines.
    Budgetary and other resource constraints.
    Is training voluntary or mandatory?
    How have training programs been evaluated in the past?
    Who conducts training?

3. METHODS AND SOURCES. Data basically can be collected in the following four ways. We can talk to people, observe their actions, assess their capabilities, and review the materials they have produced. Talking to people is often the first step in any Needs Assessment plan. Data may be gathered from both inside and outside of the organization to gain background data at the beginning of an analysis. Feedback from individuals is also used to validate and expand upon information initially collected, and it can be used to clarify outstanding questions while finalizing the analysis of the data. This approach can consist of individual interviews or group meetings. In turn, data may be gathered with a structured protocol of predetermined procedures and questions or by a more flexible approach that is directed by the interviewee, and the topics that may evolve during the discussion. Information can be solicited from recognized experts, decision-makers, and individuals who are most knowledgeable about the problem. It is important to talk to personnel who actually perform the tasks that are associated with the problem area to see if their perspective is similar to those at the top. It is also important to recognize that no problem exists in isolation and so it is important to gain feedback from those that are affected by the job incumbent, such as supervisors, subordinates and, if appropriate, clients.

Finally it is useful to consult other organizations, when possible, which may have had similar difficulties to discover how they overcame their problems. Observations yield data about day-to-day current operations. It is possible to observe job performances and group dynamics, and gain a feel for the organizational culture and work climate. Data are obtained from directly watching behavior and by interacting with the work force. It is expecially useful to investigate possible communication problems, inefficient use of time, resources, and personnel, and to identify existing and potential conflicts between management and staff. Observations are especially useful to record the actual steps, sequence, and components associated with a given task or set of job responsibilities. What people actually do can be compared with their job descriptions to assess and reevaluate the appropriateness, and efficiency of current standards and procedures can be analyzed and documented. Performance assessments produce base-line data on how jobs

Needs Assessment

are being currently performed. They generally consist of paper and pencil tests or performance demonstrations and can be used as diagnostic tools to identify current levels of knowledge and skills possessed by job incumbents. Data can be compared to predetermined job descriptions to determine where performance deficiencies may lie. Results are then used to identify needed training topics and to select personnel who will benefit the most from the training. The final source and method for collecting data is to review materials produced by the organization. Organizations generally produce three types of materials: plans, records, and end-products in the form of goods or services to be marketed or offered to the public. Materials that are generally reviewed include business plans, missions statements, productivity, and financial records, status reports, work samples, performance appraisals and management requests, policy handbooks, grievance files, exit interviews, organizational structure charts, and career development. Materials from existing training designed to teach people how to do their jobs should also be reviewed to see what type of instruction personnel have had in the past. The following presents advantages and limitations with for some of the more commonly used data collection methods:

Interviews. Advantages: Reveals feelings, causes, and possible solutions to problems as well as facts. Good opportunity to build rapport. Possible to receive additional information in the form of non-verbal information. Possible to probe ambiguous responses that yield unexpected leads. Presents opportunity to clarify expectations and assumptions about the process for the analyst and those who are contributing to the study. Affords maximum opportunity for free expression.

Disadvantages: Results may be difficult to quantify and may be affected by interviewer biases. Interviews are time consuming and are thus efficient only with a small number of individuals. Unskilled interviewers may make clients feel self-conscious or uncomfortable. In some cases interviewers may become too involved with client's problems and turn the interview into a counseling session.

Structured Groups. Advantages: Same as for interviews. Permits synthesis of different viewpoints. Promotes general understanding and agreement.

Disadvantages: Time consuming. May be difficult to schedule desired individuals for the specific meeting time. Some techniques, such as the Delphi, require interviewers to be fairly familiar with the content to facilitate a meeting.

Observations. Advantages: Unobtrusive; minimizes disruptions of the work process and activity. Yields on-the-job reports and thus provides direct contact with situations. Inexpensive if observation is informal and unstructured. Gives an idea of a typical work-day.

Disadvantages: Difficult to record data and to observe large numbers of people. Employees may feel uncomfortable. Employees may alter their behavior if they know they are being observed. Can be time-consuming. This is especially true of jobs that are part of cyclical operations that take place on a weekly, monthly, or yearly basis. Critical points may be forgotten if observers record notes after leaving the work setting. Data analysis is subject to what the observer chose to observe and record. Data analyses can be influenced by observer biases.

Questionnaires. Advantages: Can reach many people that may be geographically dispersed in a short amount of time. Yield data that can easily be tabulated and reported without difficulty. Are relatively inexpensive. Give an opportunity of expression without fear of embarrassment. No special training is necessary to administer questionnaires or tabulate results.

Disadvantages: May not ask the most effective questions and thus miss the point of the analyses. Low return rates and inappropriate responses hinder accuracy. Requires development by individuals skilled in evaluation to ensure validity and reliability of instruments. Provides little opportunity for free expression and thus limits ability to clarify and expand on responses.

Review of records, reports, and work samples. Advantages: Unobtrusive; analysts work on their own. This provides clues to trouble spots. Provides direct data on current plans and operations.

Disadvantages: May be difficult to access materials. Risks misinterpretation by unskilled analyst. Patterns and trends may be difficult to interpret from data. Analysis may require technical understanding of the material. Does not show cause and effect relationships.

Needs Assessment

Tests. Advantages: Can diagnose specific areas of deficiency. Results are easy to compare and report. Can be used to select personnel who will benefit the most from training. Can save training time and resources by screening out those individuals who do not need to be trained. Data can be compared with end of the course results to identify individual increases in knowledge and skills.

Disadvantages: Tests must be validated and developed by analysts skilled in the content as well as test development. May be time consuming to take and administer. Results may give clues but not conclusive evidence; job performance may be a better indicator. Personnel may be afraid of taking tests and therefore do not show true capabilities. Tests must be validated for this specific work situation. Tests developed outside of the work setting may be invalid for the specific situation.

4. SAMPLE SIZE. Decisions about how many data sources to use are always difficult. The desire is to minimize costs by using as small a number as possible but not so small that it would unduly jeopardize the validity of findings. Sampling is employed when it is not practical to gather data from every individual or document associated with the problem. A sample is a representative subset of a total group used to draw inferences about the entire group had they participated in the study. Samples that are unnecessarily large waste time, money, and personnel, while samples that are too small may not yield usable and accurate data. This conflict may first be analyzed by referencing size to a given stage in the study. To link the idea of stage and size it is helpful to conceptualize a study as progressing from small to large to small in terms of the number of needed data sources. The following diagram indicates the relationship between the stage and the size of the required sample.

STAGE	SAMPLE SIZE
Key Informant	Small
Validation	Large
Clarification	Small

As the purpose of the key informant stage is to explore issues, only a small number of data sources need be employed. Conversely, data need to be collected on a larger scale in the validation stage. In the validation stage issues associated with reliability are of greater concern, and so a larger sample is necessary. In turn, reliability is less important an issue when the purpose is to clarify and expand upon outstanding questions from the first stages in the study.

In addition to the study's stage, there are several other factors derived from evaluation and educational research that should be considered in making a decision about sample size. Additional variables that are useful to consider in conducting needs assessments are population size, population variability, sampling method, and desired precision (Smith, 1980). The first variable is population size. It is generally accepted that as population size increases so should sample size. The relationship, however, is not linear. As a rule, samples should represent a larger percentage of small populations than larger ones. The second issue is variability. The more people are likely to vary in their performances, the greater the sample size needs to be. A larger sample size is needed to account for differences among respondents in order to yield a more accurate estimate of average performances. This type of sampling method can also affect size. This is especially true in the case of stratified random sampling. This approach reviews data from several different groups that may vary on a critical question. For example, responses might be solicited from staff, managers, and directors on a given issue to see if and how their perspectives might differ. This requires fewer respondents than a non-stratified approach. Finally, the more important the project, the greater the need for precise, accurate results. Precision can be enhanced by increasing sample size. The greater the sample size, the more likely it will resemble the entire target audience.

Field Procedures

In conducting a needs assessment, it is important to establish a systematic way for data collection teams to work together. Use

Needs Assessment

of a number of field workers can present problems with reliability and validity of data. The data thus needs to be collected in a consistent manner to decrease the possibility of biased findings. A consistent approach is associated with the notion of equivalent conditions (Shavelson, 1981). By collecting data under equivalent conditions, the analysis can rule out the possibility that the findings were contaminated or influenced by the situation under which the evaluation took place. Standardized procedures need to be established for how questions will be asked, observations and performance assessments will be administered, and how documents will be reviewed. In turn, findings should be documented and reported in a consistent manner. Finally, routine debriefings should be standardized and established to ensure that information is shared on an on-going basis.

Data Analysis

In analyzing data, the goal is to interpret and categorize data into groups of training needs that the training program should address. This part of the Needs Assessment process is designed to do the following:
- Analyze and synthesize data.
- List needs based on identified discrepancies.
- Determine reasons for discrepancies.
- Determine which needs can be addressed by training.
- Prioritize identified training needs.
- Describe characteristics of the intended target audience(s).

Identifying Training Needs

Once the data are collected they are analyzed to first determine what performances are essential for an organization to reach present or future goals. Current performance levels are then assessed to determine if differences exist between the current and desired performances. Identified differences or discrepancies are next reviewed to determine what jobs and personnel are most associated with the discrepancies. The capabilities of job holders are first assessed to determine if performance discrepancies exist simply because personnel do not know how to do their job. Organizational factors are also reviewed to see if environmental factors impact their ability to do a good job, such as not enough time or unreasonable expectations. Additionally, it is important to review the performances of other personnel who may affect the job holders' ability to perform, such as supervisors, subordinates, and co-workers. For example, if some secretaries are diagnosed as inefficient workers, the fault may not be their own. The problem may lie with their bosses, who do not know how to properly give directions and manage their time and output. In this case, the real problem lies with management, who need training on how to better manage staff.

Performance Analysis

It is important to remember that training is not a panacea for all of an organization's problems. Performance discrepancies may not only be attributed to a lack of knowledge and skills; they may also be needs associated with a lack of motivation, environmental and other organizational problems. In order to uncover the underlying causes of identified needs, a performance analysis should be conducted. The intent of performance analysis is to identify those needs that are caused by a lack of knowledge and skill and which, therefore, can be addressed by training. Procedures for conducting a performance analysis are presented in Chapter 5.

Often more than one type of trainee will require some kind of training to solve a given problem. For example, the introduction of a new methodology for installing office automation systems may require different types of training for various members of an experienced project team. Staff may require a highly technical type of training to perform the detailed aspects of the job, while managers need both technical knowledge as well as the conceptual understanding to make needed adjustments for the specifics of a given project. In some cases it may be necessary to train one group of trainees before training others. In the above example it will be more effective to train managers first so that they are able to support and guide more junior members of the team.

Prioritizing Training Needs

Needs assessment typically uncovers multiple training needs for multiple groups of people. Training curricula are then developed for each group of potential trainees. For example, Group A may require training on topics X, Y, and Z. Group B may require training on topics P, Q, and Z. Group C may require training on topics D, E, F, and Z. Few organizations are able to address all training needs at once. They are often limited by budget, time, and lack of other resources needed to design and deliver the training as well as the limited availability of the trainees themselves. In addition one group of trainees may need to begin their training before another. For example, does it make sense to train school teachers on the benefits of a new teaching approach prior to administrators, if administrative support is needed to ensure the availability of teachers for training activities and to provide a positive atmosphere for classroom implementation? A means for prioritizing training needs by group and for a given group must be determined to ensure the most efficient response to performance problems. Several options are available for prioritizing needs. The most obvious is to ask management what needs they want to address. Such decisions are often based on vested interests that have been considered in light of existing constraints and resources. A second option is to simply select the topic(s) that are most in demand by the most number of target audience(s). In the example that described the course needs of groups A, B, and C, topic Z was required by all groups. This approach has merit in that all groups possess the same level of knowledge about the topic. Otherwise, different levels of training may need to be designed for beginning, intermediate, and advanced groups. Another possibility is to identify a stratified sample to individuals (management as well as workers) and have them select the priorities from the array of documented needs (Kaufman, 1976). Such a sorting might be made on the basis of two criteria, both generally judgmental:

    1. What does it cost to meet the need?
    2. What does it cost to ignore the need?

Individual sortings of needs can then be compared to determine consensus. Any major differences can be negotiated among those who participated. While complete consensus is almost never attained, a close proximity can be achieved. An additional option is to break down performances into specific tasks and rank them on the following points (Tracey, Flynn, and Legere, 1970):

1. Percentage of people performing the task.
2. Percentage of time spent to perform the task.
3. Probable consequences of inadequate performance.
4. Amount of time that is tolerated between initiating the cue and the actual performance.
5. Is the performance performed frequently enough to require training?
6. How much time is required to learn the task?
7. How great is the probability of a deficient performance?
8. How much time will elapse before an individual is assigned to the job and when he is expected to perform the task?

Summary

This section discussed issues that need to be accounted for in developing a needs assessment plan. In general, needs assessments are not research in the classical sense. However, basic principles designed to minimize the risk of untrue findings are considered and implemented in a practical manner. These guidelines include a need to examine the problem from a number of perspectives using a combination of data sources, to build in redundancy, gather data in a consistent manner and view the problem from both a broad as well as an in-depth perspective. Finally, the need for organizational input was emphasized to ensure commitment to the study and to better permit the gathering of relevant data in a cost-effective way. A systematic decision-making approach is suggested to determine specific guidelines for any study. The degree to which the problem is presently understood can help determine the type of data that are needed at any point in the study, which in turn drives decisions about what methods and sources need to be used to gather data. Finally, decisions about how large a sample to employ are determined on the basis of the study's stage and other basic research principles such as population size, population variability, sampling method, and desired precision. To enhance the likelihood of gathering data that are both reliable and valid, it is also important to establish systematic procedures for collecting and documenting the data.

Once the data are gathered they need to be interpreted and

categorized to decide which needs can actually be solved through training. A performance analysis approach can be used to factor out what needs should be addressed first by the organization before a training program can be effective. In some cases training is not the solution at all. Instead, the problem requires a change in organizational policy, standards, and/or procedures. The final step in conducting any needs assessment is to prioritize identified training needs. Few organizations are able to both design and deliver all required training for all designated personnel groups at once. Several options exist for prioritizing needs such as asking management to designate their preferences, using a type of sorting technique with a stratified sample of all those who participated in the study, or to rank performance tasks according to some predetermined criteria such as the listing established by Tracey *et al.* Finally, it is important to emphasize that no two needs assessments are alike. Decisions and plans will vary with the nature of the problem and the organization.

Chapter 5

Performance Analysis

Most of the training conducted in organizations is in response to a perceived need regarding the performance of some employees. Indeed, training is often offered as a solution, if not *the* solution, to many of the needs that are identified in organizations. If the sales force is not meeting projections, then they are provided with additional training. If workers on an assembly line are producing products with too many defects, then the workers are given additional training. If the forms completed by the clerical staff have too many errors, then the staff is given additional training in how to complete the forms. Time and time again, training is offered as the solution to a problem of human performance.

Appropriate Use of Training

Certainly training can and does contribute to improved performance in an organization. But can additional training solve all the problems identified by a needs assessment in an organization? Probably not. If the problem is a result of employees' lack of knowledge and skill that is required on their jobs, then it is reasonable to expect that training directed at that particular missing knowledge or skill will result in improved job performance. When a worker doesn't know how to successfully complete part of a task required on the job, then training on that task should lead to improved job performance. But what about the disgruntled worker who may know how to perform the tasks on the job but who feels underpaid or unappreciated and consequently does not try to perform the job adequately? This worker's performance might be as poor as the worker who really didn't know how to do his or her job. While the disgruntled worker may make as many errors as an unskilled worker, will additional training enable him or her to improve the job performance? Probably not. The performance does not suffer because of a lack of knowledge or skill; it suffers because of an attitude or lack of motivation to do a *good* job. A problem may also arise from the job environment itself. If the workers do not have proper tools, sufficient time, or if the procedures they are required to follow are inadequate, then job performance will likely suffer. Training rarely, if ever, compensates for these.

When training is successful. Providing training to employees may or may not result in a reduction in the need that was identified. Whether training is helpful in reducing a need depends, to a great extent, on the causes of the problem that gave rise to the need. If the need results from a lack of motivation or inadequate equipment or procedures, then additional training should not be expected to significantly reduce the problem. However, if the need is caused by a lack of knowledge or skill on part of an employee, then providing that employee with additional training can reduce or eliminate the need.

This chapter will present a formal procedure for analyzing needs to determine their probable causes and to select appropriate solutions. While the overall purpose of this book is to enable people to design effective training programs, it is equally important to learn when not to rely on training. That is, while most of this book deals with HOW to design training programs, this chapter deals with WHEN to use training as a solution to performance problems in organizations.

Performance Analysis

The purpose of performance analysis is to aid in: (1) determining probable causes of the needs, and (2) suggesting solutions appropriate to the problem. Performance analysis is appropriate when you have identified some needs within an organization that are limiting the performance of employees within that organization. Before jumping headfirst into implementing the latest "solution" that is supposed to cure any of an organization's ailments,

Performance Analysis 69

take the time to conduct a performance analysis. Such an analysis can help pinpoint the problem and direct you towards the appropriate solution(s).

The failure to conduct a performance analysis may cause you to direct resources towards a solution that is not appropriate. In absence of a careful performance analysis, you won't have the criteria to judge the adequacy of possible solutions or to select the most appropriate solution to implement. Fads and gimmicks flourish in such situations. Unfortunately, many organizations invest in additional training or more sophisticated training at these times. Should it turn out that, by luck, the problem that gave rise to the need was caused by inadequate knowledge and skill on part of the employees and that luckily again the additional training was directed towards increasing this specific knowledge or skill, then the training may well reduce the perceived problem. Of course, if the causes of the problem were something other than a lack of knowledge or skill, then additional training is not likely to reduce the incidence of the problem. It is like prescribing aspirin not for a headache but to mend a bone fracture. A carefully conducted performance analysis will increase the likelihood of finding a specific solution to a performance problem that will reduce the incidence of that particular problem.

Procedures for performance analysis. There are five steps in performance analysis that are based on a model recommended by Deterline (1974). These steps are shown in Figure 5.1.

Examine needs. The first step in analyzing a performance problem is to examine the need, describing it as clearly as possible. An appropriately conducted needs assessment is the necessary starting point for performance analysis. Often first attempts at needs assessment consist largely of description of the current state rather than the discrepancy between the current or actual state and the desired or ideal state. It is much easier to discover that the error rate on a certain product has increased from three percent to five percent than it is to discover whether this is a problem or what caused this increase. Likewise it is easier to discover that the number of forms from the accounting section that had to be re-done by a supervisor has increased by forty

---

Figure 5.1 Performance Analysis.

1. **Examine Needs**

2. **Identify Causes**

3. **Identify Potential Solutions**

4. **Select Appropriate Solution**

5. **Plan For Implementation**

---

percent than it is to discover the underlying problem that might be causing this increase.

For example, through needs assessment you may document that fifteen percent of the new applications taken by your marketing division for some service you provide had errors in them. This describes the current state of affairs. But does this provide ample evidence that a need exists? Well, fifteen percent might seem high, but what are the industry norms? If other, similar companies had an error rate in excess of twenty-five percent, then you might be pleased with your relatively "low" error rate!

Since the purpose of performance analysis is to identify probable causes of the problems that exist in an organization and to suggest possible solutions, performance analysis requires that an appropriate needs assessment was completed. The needs that were identified and justified will be analyzed.

Identify causes. The second step in performance analysis is to explore the underlying causal factors that may be contributing

Performance Analysis 71

to the performance deficiency identified in the needs assessment. There are, of course, many reasons why things may go awry in an organization, causing problems with the performance of the employees. The various causes of performance problems are grouped into three general categories: (1) knowledge/skill factors, (2) organizational/environment factors, and (3) motivational/attitude factors.

The purpose of this second step in performance analysis is to determine which of these three factors is most likely causing the performance deficiency. This search for underlying causes is the most analytical step in performance analysis; it is also the most difficult step in performance analysis.

The basic approach to identifying the probable underlying causes is to ask a series of questions that explore each of the three categories of causes. To determine if the cause might be a lack of knowledge or skill, you might ask "Do the employees have sufficient knowledge and skill for the tasks?" To determine if the causes are in the area of organizational or environmental factors, you might ask "Does the organization or the work environment interfere with satisfactory job performance?" Finally to determine if the need is a result of poor motivation or attitudes on part of the employees, you might ask "Could they do the tasks if they really wanted to?" While these questions will not pinpoint the specific cause, they will direct your attention to the major category of the causes and allow you to then focus on more specific causal factors.

Identify potential solutions. The third step in performance analysis is to identify alternative solutions for resolving the need. It is in this step that you select potential solutions to the need that are appropriate for that need. This matching of the solution to the problem is the heart of performance analysis. If the cause is a lack of motivation, then the solution must address the motivation and might include increased rewards or recognition. If the problem is a lack of knowledge, then the solution would likely include additional training. It is important that the solution match the problem. Increased recognition is not likely to solve the problem if it is due to a lack of knowledge. Likewise, additional training will not likely remedy a problem caused by a lack of motivation. The solution must be responsive to the cause of the performance problem.

Select appropriate solution. The fourth step in performance analysis is to select the solution to be implemented. In the third step, alternative solutions to reducing the need were identified; in the fourth step, you determine which specific solution to implement. There may be several specific solutions that seem to "fit" the problem. At this point in performance analysis a choice must be made as to which solution seems the best for the given circumstances. Of course, it is possible to select more than one solution to implement.

Plan for implementation. The fifth, and final, step in performance analysis is to plan the implementation strategy and then implement the selected solution. By working through all the steps in performance analysis, you increase the probability of resolving the performance problem that underlies the need by selecting and implementing an effective solution to the problem.

Using Performance Analysis

Now that we have seen an overview of the performance analysis procedure, let's focus more closely on the specific steps. We would begin performance analysis whenever there was an identified need related to the job performance of some employees. As such, performance analysis begins by exploring the need. When error rates increase, customers become more disgruntled, or sales decrease, you might expect that there is a performance problem underlying this state of affairs. Thus, you begin to collect indicators about the problem. How much have sales declined? What specific product has had the greatest decline in sales? Which region, district, or sales person has experienced the decline? Performance analysis requires that data about the incidence of the need be collected prior to making the determination about the severity of the need and its probable causes.

Isolating causes. We know through the needs assessment that there is a problem or discrepancy sufficient enough to justify taking some action. In this step of the performance analysis we seek to determine if the performance deficiency is most likely

Performance Analysis 73

a result of a lack of knowledge, some organizational factor, or a lack of motivation. We can explore each causal factor in turn by asking a series of questions.

Knowledge/skill factors. We first ask "Does the employee have adequate knowledge or skills for the tasks?" If the answer is negative, then we explore the knowledge/skill component in more depth. We then ask other questions that are designed to help determine whether the performance deficiency arises from a lack of knowledge or skills on part of the employees. Typical questions are shown below:

- Has the employee ever performed the task adequately?
- Has the employee received specific training for the task?
- Does the employee often perform the task?
- Does the employee always perform the task wrongly?
- Does the employee receive feedback on his/her performance?

From asking questions similar to these, you determine whether the probable cause of the performance deficiency is a lack of knowledge/skill.

Organizational factors. If the employee has adequately performed the task in the past or has received training directed at the specific task, it might be reasonable to assume that the performance deficiency is not due to a lack of knowledge or skill. We would then ask "Does something in the job environment interfere with the performance of the job tasks?" If the answer is positive, we would then explore factors in the job environment that may be causing the performance deficiency. We could ask questions such as:

- Does the employee have adequate tools or information?
- Is the employee frequently interrupted?
- Are job standards communicated?
- Can the employee describe how to perform his/her tasks?
- Have employees complained about working conditions?
- Are there rules that conflict with task completion?
- Are job standards too high?
- Are there problems common to many employees?

By asking questions similar to these we should be able to determine whether the cause of the performance problem is some factor in the job environment.

Motivation/attitude factors. If the answers to the above questions indicate that nothing in the job environment is causing a performance deficiency, we would then ask "Is the employee motivated to perform well?" It might be possible that the performance deficiency results from a poor attitude or lack of motivation on part of the employee. To determine if this is the most likely causal area we could ask questions such as:

- Could the employee perform the task if his/her life depended on it?
- Is there something he/she especially dislikes about the task?
- Is there anything about the task that frustrates the employee?
- Does the employee receive praise for an adequate performance?
- Are there any negative consequences associated with a desirable performance?
- Are there somehow certain advantages (to the employee) associated with a poor performance?
- Is the employee informed about the quality of the performance?

Depending on the answers to these questions, there may or may not be reasons to expect that the cause of the performance deficiency lies in the motivation/attitude area. It may be that the employee could perform adequately if his or her life depended on it or if he or she really wanted to do so. Thus, the cause of the problem is not a lack of knowledge or some factor in the job environment. Rather the cause would be associated with a lack of motivation or a poor attitude about the job.

An example. Let's return to the example of declining sales in an automobile dealership. Let's assume that our sales were down considerably when compared with industry norms; we have an actual need. Further, let's assume that our sales staff had been selling an acceptable number of automobiles up to a few months ago. All our sales staff had received training upon employment and had seemed successful in using the sales techniques they were taught. From this information we should assume that the cause of

Performance Analysis 75

our performance problem is not a lack of knowledge or skill. The sales staff performed adequately in the past, all had received training, and all had demonstrated that they could apply the skills in which they were trained. Thus, we would focus our investigation of causal factors on another area. We might consider whether the causes were in the motivation/attitude area. Certainly, we believe the sales staff could perform adequately if they really wanted to, if their lives depended upon it. In the past their work had been satisfactory. We now focus in on attitudes a bit more. We recall that several sales persons have complained about their income, especially their commissions. A couple of our better sales persons quit and went to work for another dealership. We might talk with a few sales persons to gain their perspective on this potential attitudinal problem. We discover that most of their salary comes from the base rate; very little is added by commissions on actual sales. That is, they make about the same amount whether they sell a car or not. There has been no recognition of the top sales person each month or any other incentives for hard work. As one sales person said, "Why try? I can loaf around and make as much as if I busted my chops." From this information we could conclude that the cause of their performance problem was in the motivation/attitude area.

Explore solutions. Once we have identified the causal factors, then we explore alternative solutions to the performance problem. An organization has many options for dealing with performance problems among its employees. These options can be grouped into the same three categories as the causes were grouped: (1) knowledge/skill factors, (2) organization/environment factors, and (3) motivation/attitude factors. As previously indicated, a main purpose for doing a performance analysis is to match the solution to a performance problem with the causes of that problem. So, if our analysis indicated that the cause related to a motivational or attitudinal factor, we would select a solution that was appropriate for such a problem. In this case we might provide more incentives for a good performance. In the example of the automobile dealership, we might try to increase the commission rate. We could also provide some bonus to the top sales person at the end of each quarter or month, perhaps a weekend at a beach resort. Perhaps the sales staff feel left out of decision-making that affects their work. We might be able to arrange for sales personnel to have a voice in certain decisions. Maybe they could plan their own hours rather than being assigned work times without regard to their wishes. Perhaps they could assist in determining what types and how many new cars to order from the factory. While there are no guaranteed solutions, there are some solutions that are more appropriate in light of a given problem and its causes. Performance analysis seeks to discover these more appropriate solutions.

Solutions for knowledge/skill problem. There are several potential solutions to a performance problem when the causal factors indicate a lack of knowledge or skills. The most common solutions to a knowledge/skill deficiency are:

- train the employee
- replace the employee
- develop a job aid
- redesign the job

Certainly the most often selected solution to a performance problem caused by inadequate knowledge or skill is to provide additional training to the employee whose performance suffers. This training could take the form of traditional classroom instruction or might involve more informal on-the-job training. The employees might also be encouraged to enroll in some course offered elsewhere, perhaps by a technical institute or nearby university. There are many options for who provides the training, where the training takes place, what type of training is offered, and the like. At this point it is sufficient to indicate that provision for training is one potential solution to performance problems that arise from a lack of knowledge or skill on part of employees.

Another option to compensate for a lack of knowledge or skill on part of an employee could be to replace that employee. At times it may be more appropriate to reassign an employee rather than provide extensive training. It might be a case of having the wrong person in a certain position. Additional training may be a less effective solution than getting a new person on the job. While reassigning or terminating an employee should not be done lightly, this may be the best option in certain circumstances. This decision requires informed judgment. You must weigh the probable bene-

Performance Analysis 77

fits and costs of additional training versus getting a new employee in that job.

A third alternative solution to a performance problem caused by inadequate knowledge or skills might be to develop a job aid. If a great deal of specific information is required on a job and the job incumbent can't seem to recall all of the information as it is needed, then you might develop a job aid that can serve as a quick reference. For example, some sophisticated computer programs have many different commands that the user must enter to cause certain actions. A user could spend many hours trying to memorize all of the commands for a software product. However, the user would not likely recall many of the commands if he or she did not continue to use them frequently. A preferred option taken by some software companies is to develop a sheet to fit over the computer keyboard that contains a summary of the various commands. That way, the person using the program doesn't have to remember each and every command; he or she can quickly look it up on the reference sheet. Such job aids can facilitate work by expanding a person's ability to perform adequately. Job aids can reduce the amount of information that must be memorized and thus reduce the problems that arise when this information wasn't remembered.

A fourth alternative solution to a performance problem caused by inadequate knowledge or skill is to redesign the job. If some employees don't have the skill required to perform some tasks required in their jobs, then see if the tasks can be done differently to take advantage of skills they *do* have. Perhaps you can re-do some procedures that are followed. Maybe some aspect of the task could be automated. Perhaps the task could be subcontracted out and completed elsewhere. Whether the job could be redesigned depends upon the specific nature of the job itself. Thus, it is not possible to give specific advice on redesigning jobs. Rather we want to mention that when there are performance problems caused by a lack of knowledge or skill, you should explore the possibility of redesigning the job to eliminate the requirement for this knowledge or skill.

Solutions for job environment problems. If the analysis shows that the performance problem is caused by organizational or environmental factors, then different solutions are required. These

include:
- job design
- change in policies or procedures
- change in management practices

If the performance problem was caused by some obstruction in the workplace, then that obstruction must be identified and removed. First consider the design of the job itself. Do the employees have adequate tools and equipment for the tasks? Are necessary tools and equipment well maintained? Is sufficient workspace available, and is this workspace conducive to doing their tasks? Are the tasks arranged in a logical, easy-flowing manner from the *workers'* points of view? Are necessary support persons available to assist the employees? In essence you must examine the job to determine if it is do-able as it is currently configured. Interview employees to discover problems they are having with their jobs. Get their suggestions for changes in the way the work is done.

Solutions for motivation/attitude factors. If the performance analysis indicates that the most probable cause of the performance problem is poor motivation or "bad" attitudes, then solutions appropriate for these causes must be explored. These solutions include:

—increase the incentives for acceptable performance;
—increase feedback about job performance;
—remove any negative consequences for acceptable performance;
—remove reinforcement for poor performance; and
—reduce unpleasant aspects of the job.

It may be that the incentives for performing at an acceptable level are too few or too weak. If this is the case, one way to increase the quality of the performance is to increase the incentives. Consider putting the sales staff on commissions for some of their pay or increasing the rate of the commissions they currently get. Let the workforce share in the savings from fewer errors if their performance improves. In essence, make sure that the rewards for good performance get back to the people that performed well. Another approach is to increase the feedback that employees get about their job performance. Personal feedback is a strong positive reinforcer for most people. Let them know how they are doing, and let them know that their doing a good job is important to you. Find and remove any negative consequences associated with doing the job adequately. Make sure that doing the job well isn't more punishing than doing it poorly. For example, if the job is inspecting work products for defects and everyone gets a reduction in pay for any defects that are found, the inspectors will likely be punished by their peers for doing their jobs well. Furthermore, if everyone, including the inspectors, has a salary reduction for each error found, then finding errors is directly punishing to the inspectors. Be certain that a poor performance is not rewarded. If you tie a person's evaluations and pay to the rate at which he or she performs, regardless of the quality of the performance, expect a lot of poor work! It is important to make sure that you are not "paying off" on poor performance equally with good performance. It all comes back to incentives. What are the incentives for doing a job well? Are they more than for a job poorly done? If not, then no amount of additional training is likely to improve the quality of the work and reduce the performance problem. Finally, you should identify and remove, if possible, any unpleasant aspects of the job. Interview or survey employees to find out the parts of their jobs that they don't like or enjoy. Find out what "bugs" them about their jobs. Then change or modify those parts where possible. Sometimes even relatively small changes in a job can make it much more pleasant to the worker.

Summary

In summary, performance analysis follows and expands upon needs assessment by identifying and analyzing likely causes of the need in order to suggest appropriate approaches to reducing that need (Mager and Pipe, 1984). Depending on the nature of the causes of the needs, training may or may not be advisable as a solution. Some needs are not likely to be reduced by additional training. These needs are those that are caused by problems with motivation or attitudes of employees or by organizational problems that interfere with the employees' ability to work. Training, then, is not a proper solution to all needs.

Chapter 6

Task Analysis

A fundamental principle of instructional systems development is that the instruction must enable the learner to perform better once he or she has completed the instruction. This is equally true whether the person is receiving instruction to become a lawyer, a doctor, an automobile mechanic, or a diplomat. Consistent with general systems theory, you can view the instructional system as a sub-system of a larger system. In such a framework the instructional system is influenced by the larger system, i.e., the environment or organization, in which it operates. The environment provides the resources, both human and fiscal, for operating the instructional system. The larger environment also constrains the instructional system through various rules and regulations as well as through beliefs commonly held in the environment. Training, in turn, supports the needs of the larger environment.

Those learners completing the training will return to the environment, and their success depends on the extent to which the instruction they received enables them to perform successfully in that environment. Thus, successful instruction must be based on a careful assessment of the skill and knowledge that would be beneficial in the environment of the instructional system. It is for this reason that task analysis follows a careful determination of needs.

For a moment suppose that you are the training director of a major pharmaceutical firm responsible for the education and training of a large and diverse staff. One major aspect of your total program is the training of the sales staff. How do you establish the purpose of the sales training? How do you identify the content that should be transmitted during sales training? The answers to these questions are found by looking outside the training division, since the training division as a sub-system exists to provide outputs, e.g., more knowledgeable sales persons, to the larger system, in this case the pharmaceutical firm. If the training division is to be judged as successful, then the people who have received training should be better at sales than before. That a person learned more about specific drugs and sales techniques during the training is an indicator of his or her intelligence and diligence but we can't evaluate the training adequately until we observe whether this person increased his or her sales volume. He or she was in training to increase the sales, so this is what should be examined to determine the effectiveness of the training. What a person learns during training should impact what he or she actually does on the job; the knowledge and skill from training should be applied on the job.

Purpose of Task Analysis

If training is to promote enhanced job performance and contribute to an organization's mission, then the skills that one uses in performing the job should be the basis for the training. The intent is to focus the training on specific job performances. Ideally, the training should include instruction on all essential aspects of the job and include no instruction on non-essential aspects. Essential aspects are those skills necessary for performance of the job as well as the knowledge that supports job skills. Other information that might be related to the job but is not required in performance of the job—the nice-to-know information—would be excluded. When the intent is to develop a training program to prepare people within a large organization to function more effectively in their jobs, the logical starting point in deciding on the objectives for the training is an analysis of their jobs. Job-task analysis, or task analysis as it is often called, is an early step in ISD projects that forms the basis for later development of the instructional content (Branson et al., 1975). It is by focusing on careful analysis of the jobs performed in an organization that training can contribute to an organization's effectiveness.

When taken together, task analysis, needs assessment (Chapter 4), and performance analysis (Chapter 5) are the initial steps in ISD

Task Analysis

models. The ordering of these three analytical methods is important, since each is for a different purpose. Needs assessment usually occurs first and seeks to determine if there are important discrepancies between the ideal, or expected, performance of persons in the organization and their actual performance. If the discrepancies are relatively minor and not worth acting on, then there is no reason to develop a training program. However, if needs are identified and the decision is to act to alleviate the needs, performance analysis will help uncover the probable causes of the needs and will suggest strategies to reduce the needs. For those performance problems, or needs, that are found to arise because of employees' lack of knowledge and skill, task analysis procedures can be used to pinpoint the necessary knowledge and skill that will be the focus of the training program. Thus, needs assessment, performance analysis, and task analysis are complementary procedures used in the initial phase of ISD. Each serves a different purpose within ISD; all are necessary for successful development of educational and training programs.

Contributions of Task Analysis

Task analysis methods were pioneered in the military services and have found their way into training programs in large corporations. Task analyses offer several distinct benefits within ISD models. It is the task analysis that sets the direction for the training. In the absence of task analysis, training programs may include much content that is not related to job performance and may fail to include essential content. The results from such training programs are not acceptable in most organizations. Employees who complete such training are not capable of performing all the duties of their jobs because of hit-or-miss training programs that are not based on task analysis data. Such training programs are inefficient in that training time and resources are wasted in providing instruction that is not related to the job duties of persons receiving the training. Thus training programs that are not driven by task analyses are unnecessarily costly and inefficient. Large organizations must be able to ensure that their training programs are effective in enabling people to perform on their jobs and that they do so with a minimum expenditure of resources. Organizations invest in training expecting, often demanding, a return on their investment. Each day an employee is in training is a day that the employee costs the company rather than contributes to the production of goods or the delivery of services for the company. Training is costly; training not guided by task analyses is unduly costly and uneven in its effects. Most organizations in a competitive economy cannot afford training programs that don't use some form of task analysis to guide their development. Training programs in public agencies likewise must be productive because of scarce resources available for training and pressures on the training programs to turn out people who can perform adequately on their jobs. For these reasons task analysis is a common procedure in instructional systems development models.

Approaches to Task Analysis

Job task analysis is a procedure that involves collecting data about what people do when performing a job. Most task analyses begin by dividing a job into smaller parts of the job as shown in Figure 6.1. These smaller parts of a job are usually called duties.

Each duty may be divided into smaller units called tasks. While a person may spend considerable time performing a job, a given duty would not occupy many weeks. Rather, a duty is at a level that can be completed in a smaller unit of time than a job. Duties have a beginning and ending time associated with them. Often a duty may last for several hours, perhaps days, but rarely weeks. As the smallest unit of a job, a job task is accomplished typically in hours. A complete job task analysis would include an indication of all the duties and tasks that make up that job (McCormick, 1979).

Once a job has been broken down into specific duties and tasks, additional data about these duties and tasks are collected. These data may be listings or *inventories* of the job duties, *descriptions* of each job duty and task, information about each job task to be used in *selecting* tasks for training, the *sequence* for performing or learning the tasks, or the identification of the *prerequisite content* for each job task. There are different task analysis models for each of these different approaches (Jonassen and Hannum, 1986).

Figure 6.1 Job Breakdown.

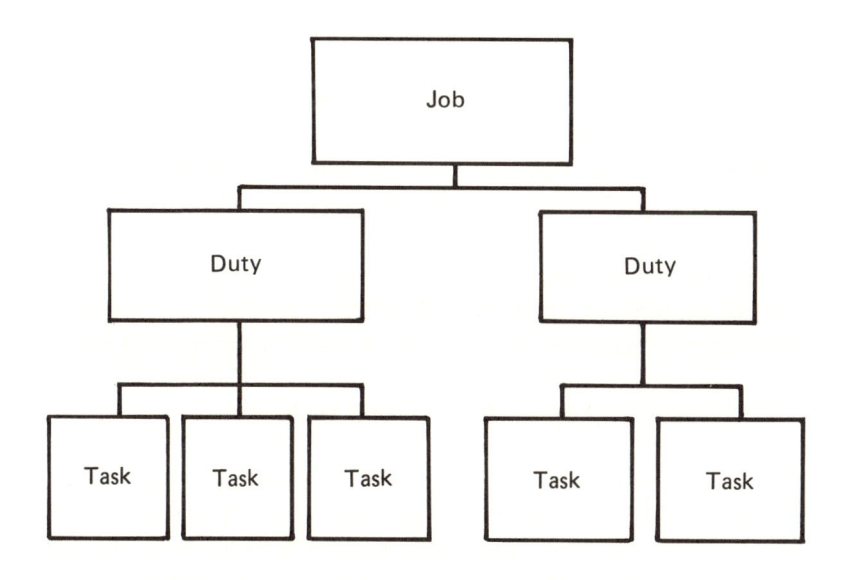

Data necessary for task listings may be collected from job incumbents through observation, surveys, or interviews. One approach to collecting data to list job tasks involves observing incumbents perform a job and recording the specific duties and tasks they did. In this manner the job analyst can develop a listing of the various tasks people do when performing a job. An alternative is to ask job incumbents to indicate the tasks they do when performing a job. These data may be collected through mailed surveys or through interviews.

While the intent of each of these approaches is the same, to collect data about the duties and tasks that make up a specific job, there are some differences in how each approach is carried out. Observing incumbents perform their jobs requires many trained observers and, as a result, is expensive to conduct. Careful observers can collect rich data about what one does on his or her job. These data may more accurately reflect what happens on a job than do reports from incumbents about what they do. However, because of the time requirements for gathering observational data, the number of people observed is more limited than in the case of surveys or interviews. There is also the concern that the incumbents changed what they normally do on their jobs because they were being watched. Thus, observational data may be somewhat suspect to bias.

More data can be collected with interviews than through observations because interviews are less time consuming. In interviews job incumbents have a chance to expand on what they do and they may include tasks that an interviewer might miss through observation alone. However, job incumbents may report what they think they should do on their jobs, not what they actually do. Job incumbents may skip over some tasks during the interview, failing to recall all that they do. Because it is a one-to-one endeavor, interviewing does require considerable time, although perhaps not as much as observation.

Surveys of job incumbents allow for easy collection of much information in a small time period and require less labor than observation or interviews. However, it is not possible to ask follow-up questions based on a person's responses or to probe further about certain tasks. So there are trade-offs in completing task listings. Selecting the data collection technique to use is a matter of judgment given the situation in which you are working.

Beyond listing the specific job duties and tasks, the job task analysis may include a *description* of these duties and tasks. These descriptions would detail what a person does and how he or she does it when performing a job task. It is one thing to indicate that a mechanic must remove and replace an alternator; this would be part of a task listing. It is different to indicate what removing and replacing an alternator involves; this would be part of a task description. A task description elaborates on the tasks in the task inventory by describing the steps or procedures followed in performing the tasks.

In some task analyses additional information about the tasks is collected to assist in *selecting* tasks for training. This information usually includes the number of persons who perform the task, the

Task Analysis

frequency with which they perform the task, how critical the task is to successful job performance, how difficult the task is to learn, how much time is spent performing the task, etc. Since most organizations rarely have the luxury of training each employee on every possible job task, decisions have to be made about which tasks to provide training on. Such task selection procedures are routinely used in military training.

The recommended *sequencing* of tasks is part of some task analysis models. This sequencing may be little more than the sequence in which the tasks are performed on the jobs. This may represent an appropriate instructional sequence for some tasks but may not be sufficient for others. The optimum sequence for learning some tasks differs from the sequence in which the tasks are routinely performed. Instructional sequences may be based on considerations other than the sequence in which the tasks are performed. For example, learning hierarchies suggest a different sequence for learning a task than the sequence in which the task is performed.

A final type of task analysis identifies the *underlying knowledge or skill* that supports task performance. This type of task analysis is more cognitively oriented, seeking to identify what a person must "know" before he or she can perform a task. This approach requires that the person analyzing a task, or set of tasks, attempts to identify what must be known before a person is capable of performing a task. Other task analysis approaches are more behavioral in their orientation, only dealing with the observable components of job performance and not attempting to examine what is going on in one's head when performing a job.

Depending on the particular needs of an organization, a different task analysis procedure would be used. Of course, several of these analytic methods might be combined and used in an ISD model. A complete task analysis procedure would list the tasks, describe them, select tasks to be included in training, identify their sequence, and describe the supporting knowledge. Whether an organization uses such a complete task analysis depends on the availability of time and resources as well as the intent of the analysis. Trade-offs are inevitable in instructional systems development and thus it is not unusual for different organizations to use slightly different task analysis procedures. What is unusual is for no task analysis procedure to be used; this clearly violates an important aspect of instructional systems development.

Conducting Task Analyses

Very large organizations and some installations in the military services have training personnel who devote full time to conducting task analyses. In other organizations task analyses are conducted by persons in the training department who perform other procedures in the overall ISD process. Whether done as a full time assignment or not, task analyses are performed in a similar manner. The first step is usually to list, or *inventory*, the tasks that make up the duties on a job. As previously mentioned, this information is usually obtained from job incumbents. Task analysts may observe several skilled employees perform a specific job and from these observations the analysts create a listing of the specific steps the job incumbents completed in performing their jobs. In some cases task analysts use videotape to record the performance of incumbents for a given job. After the performance is recorded, the analysts review the videotape to identify what the employees did in performing the job. The videotaping allows multiple viewings of the job performance rather than just a one-shot glance at each employee performing his or her job. This ensures that the person analyzing the job can be more certain about what tasks are performed. However, the act of videotaping itself may alter the performance of the employees being videotaped; they are on-stage.

Another approach to creating a task inventory is to interview rather than observe job incumbents. Job analysts can ask a small sample of employees to tell them what they do when performing a certain job. Information about the reported tasks are collected and examined for commonalities. Another approach is to collect data from more employees through a mailed survey rather than through an interview. The survey asks each employee to list the tasks he or she does on the job. This allows for the collection of data from more people but removes the opportunity for the person analyzing the job to ask for clarification or more information.

Task Analysis

Information about *task descriptions* is gained in a similar manner through observation, interview or survey. In any case the person analyzing the job attempts to go beyond the listing of tasks by describing the tasks that are performed. This description usually includes information about the cues that initiate the performance of the task, the steps or movements in the task itself, the circumstances surrounding the task performance, the use of any tools, and salient aspects of the task to which a person must attend. Some task analysts view the videotaped performance of a person completing a task with that person and have that person describe exactly what he was doing. Thus the task description will reflect the incumbents' descriptions rather than solely that of the task analysts.

Task selection information is usually obtained through survey methods so that sufficient data are gathered. Task listings are sent to many job incumbents who are asked to provide specific information for each job task. Usually they are asked to indicate how often they perform each task, how long each task takes to perform, how difficult each task is to learn, how difficult each task is to perform, and what the consequences are of performing a task poorly or not at all. By combining data from many job incumbents the task analyst can examine the tasks for a job and make determinations for which tasks training should be developed. Those tasks that are performed by the greatest number of people on a particular job are more difficult, are performed more often, and are more critical to successful job performance are selected for training.

Attempts at *task sequencing* often follow the sequences that were observed when incumbents performed their jobs. That is, the instructional sequences are patterned after the sequences performed on the job. Many training analysts consider this to be the most appropriate instructional sequence because it represents high "fidelity" with the job. This is typical of behavioral approaches to task sequencing. A cognitive approach offers alternative ways of sequencing the instruction. A distinction is made between the optimal performance sequence and the optimal instructional sequence. For example, the elaboration sequence of Reigeluth specifies a general-to-specific sequence for instructional presentations. This stands in contrast to a fixed linear sequence of actions performed on the job.

Perhaps the most complicated and ambitious aspect of task analysis is the identification of the *underlying knowledge or skill*. Gagne's work on learning hierarchies represents a common approach to identifying the prerequisites for certain performances that involve the use of concepts, rules, or problem solving. In developing a learning hierarchy you begin with the specific job skill and ask "what must a person be able to do in order to do this." Then for each prerequisite you again ask what a person must be able to do and so you proceed to identify the necessary prerequisites that lead to the ability to perform a certain task. The result of this is identification of the prerequisites.

One of the earliest approaches to identifying instructional content came from programmed instruction. First the end point (job task) was specified, then this task was broken down into a series of small steps, or successive approximations, that lead up to accomplishment of the task. Such behavioral analyses serve as the basis for identifying what must be learned and representing it as a behavioral chain.

There are other approaches to identifying the supporting content for a task that are based on a cognitive perspective. Reigeluth and Stein (1983) recommend identifying the main idea or epitome and then identifying the specific aspects of the idea. By following this approach, you would identify the supporting content that must be understood before a person could perform adequately on a job. This represents a general-to-specific structure. This seems similar to Ausubel's suggestion that instruction proceed from the general to the specific, progressively differentiating broader concepts into more specific ones.

There are various approaches to identifying the instructional content that will support certain job performance. The more complete task analyses attempt to ascertain this supporting content to serve as the basis for the training. Techniques for task listing, description, sequencing, and selection are much more standardized than the techniques for identifying the instructional content. ISD is still an emerging discipline; procedures for specifying instructional content are still evolving and improving.

Summary

Task analysis is an essential step in instructional systems development. It serves to focus and constrain subsequent steps in ISD models. There are several aspects to task analysis including task inventories, task descriptions, task selection, sequencing of instruction, and identification of the instructional content.

Chapter 7

Goals and Objectives

Training designers are like architects. They need a set of blueprints before they can begin developing materials for their training programs. This set of plans is called the training design and should specify the best way to address training needs that were identified during the front-end analysis. The design thus serves to translate performance gaps into goals and objectives that are to be accomplished by the learner.

Goals and objectives describe the purpose; that is, they describe the intended outcome of the training. They begin by broadly stating how the training will contribute to society's needs or the organization's mission or business plans and are further refined and detailed until they identify solutions as an intended performance for a given group on a given task(s) or sub-task(s).

Generally speaking, training programs begin with a macro, or big picture, type of analysis and planning. In macro level planning, training goals and objectives are first linked to society's needs or the organization's mission and are thus associated with long-term planning and outcomes. This type of analysis begins by first determining what organization(s) or department(s) within an organization is the most appropriate source to supply and manage the necessary skills and resources required to support the overall plan. The identified organization or department is then assessed to determine its present capability to carry out its role. As a result, various groups of individuals may be identified as requiring training in new skills and knowledge to expand upon or improve their present performances.

In this part of the design process, it is common for multiple target audiences to be identified who may require training in multiple subject areas. For example, a program directed by a national health department in the area of adolescent pregnancy might require some kind of training for school administrators, nurses, parents, community, and religious leaders as well as for the adolescents themselves. In turn, training topics could range from the use of contraceptive measures, to counseling techniques designed to help adolescents better understand the risks and dangers associated with early pregnancy.

A more detailed type of analysis occurs in designing the individual course and lesson. At this point, different types of analyses and skills are used as the design process is linked to specific job performances. In course design, goals and objectives are thus stated in terms of expected behavior that participants are expected to perform at the end of the instruction. Content and other instructional decisions such as sequence, testing, and the selection of methods and media are made by determining the type of skills and knowledge required to reach the terminal behavior. Courses are then subdivided into individual lessons designed to carry out one or more course level objective(s). Lesson design is further organized to incorporate instructional strategies based on theory and research to enhance learner comprehension and retention.

The Four Design Levels

To better understand how goals and objectives are used to state outcomes for both macro and micro types of analyses, it is necessary to think of the design process as occurring on different levels. The four levels are the program, curriculum, course, and individual lesson. It is helpful to think of the program and curriculum levels as "macro" design and the process used to develop courses and lessons as "micro" design. In the design of any training program these levels will constitute a sort of hierarchy. Within a given program, for example, there may be several curricula. In turn, a curriculum consists of a series of courses, and a given course will comprise a number of different lessons. Note the diagram in Figure 7.1.

Goals and Objectives 95

Figure 7.1 The Four Levels of Design.

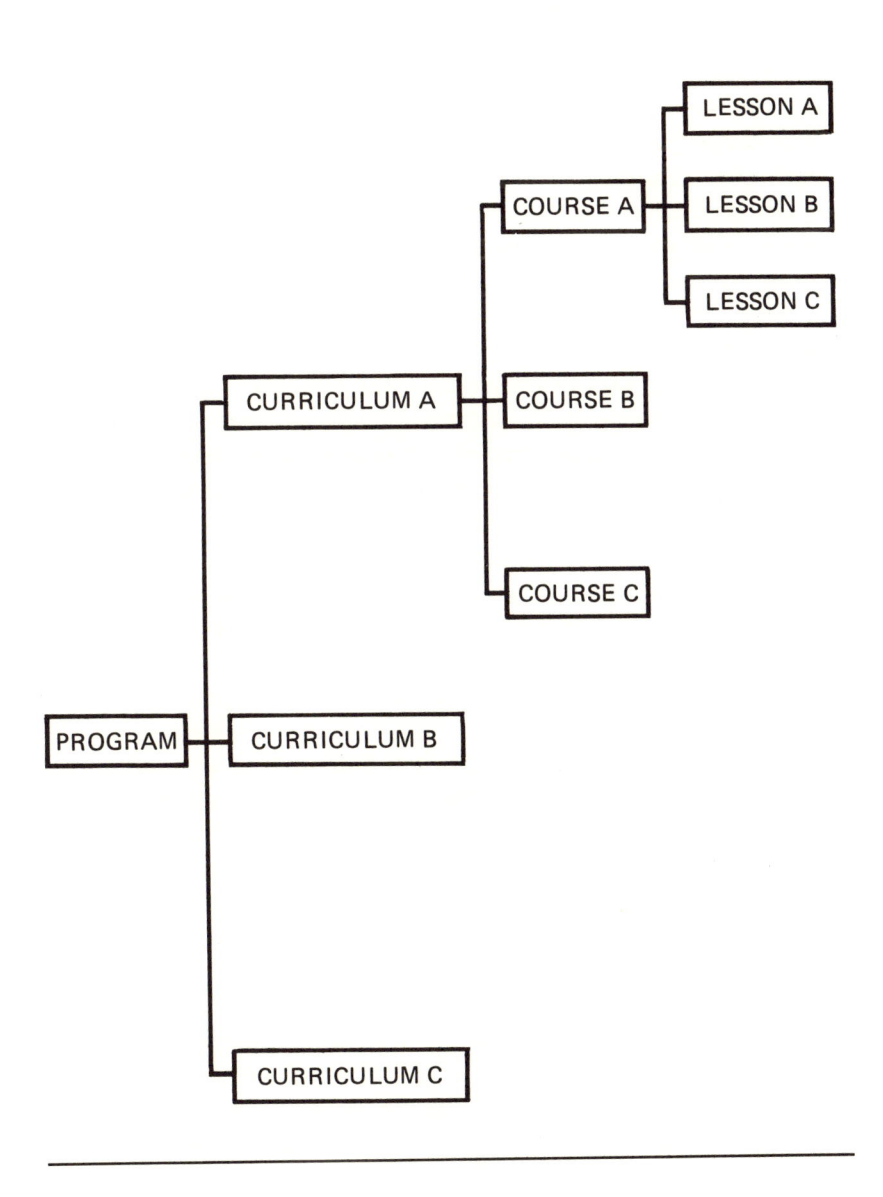

```
                                    ┌── LESSON A
                      ┌── COURSE A ─┼── LESSON B
                      │             └── LESSON C
                      │
        ┌─ CURRICULUM A ─┼── COURSE B
        │             │
        │             └── COURSE C
        │
PROGRAM ─┼─ CURRICULUM B
        │
        │
        └─ CURRICULUM C
```

Macro Design

The program level begins by analyzing the "big picture," that is, how training can contribute to solving the overall problem. As discussed in the chapter on needs assessments, problems may originate from a desire to improve general living conditions such as better health care, a change in the environment, or from an organization's inability to meet its stated goals or mission. Educational programs are often one of several components of a large-scale effort. Other components may include, for example, the need for increased manpower, expanded facilities, and the passing of new legislation. Because program level goals and objectives are generally associated with long-range planning, they may take several years to achieve.

The following is a hypothetical example of a health education program in a third-world country. In this problem, a need exists to address the high incidence of waterborne disease in rural areas. Waterborne diseases can attack all parts of a population and are considered chronically debilitating and potentially life threatening. While the parasites that cause this type of disease live primarily in small streams and ponds, they can also be transported by humans and certain animals for a period of up to 48 hours. It has been determined that villagers do not link sanitation habits with disease. For example, they do not wash their hands in boiled water before preparing food and they bathe and swim in contaminated streams and ponds. The problem is in part associated with poor water resources in rural areas, which requires funding and manpower to build new water sources and decontaminate existing ones. However, the problem is also linked to inappropriate practices and attitudes held by the villagers themselves. Villagers will need to practice better sanitation habits to protect themselves in the present and to avoid the risk of further contamination once improved water sources are available. It is feasible that this part of the problem can be addressed through a health education program.

In this example, the Ministry of Health has been designated as the most appropriate source to supply the skills and manpower required to carry out such a training program. We will describe the macro planning for this training project for both the program

Goals and Objectives 97

and curriculum levels. The first chart lists the overall program level goal and its corresponding objectives. They are the organizational roles and tasks that are required to enact the training program. Note that the program goal broadly states the educational component and how it will contribute to the reduction of waterborne illness. The objectives, on the other hand, describe what personnel within the organization must possess what capabilities to reach the program goal. As in most instances, certain personnel roles are created specifically for this project, while others are changed or expanded to improve their capability of meeting program requirements. A description of training topics for each personnel group will follow in our discussion on curricula.

GOAL: To develop an educational program that can help decrease the incidence of waterborne disease in rural areas in country X.
OBJECTIVES:
1. To institutionalize within the Ministry of Health the capacity to plan and implement health education program via the creation of a national health education unit that will affect health related attitudes and practices of rural people.
2. To institutionalize with the District Health Education unit the capacity to plan and implement the above described national health education program for rural people.
3. To institutionalize interministerial cooperation on all levels (central, district, and community).
4. To utilize existing community social networks and community based personnel in order to gather information and conduct training related to water use contact and sanitation habits.

In this example the first two objectives describe how the Ministry of Health as an organization will respond to the problem. First, the ministry will accept the responsibility for creating and administering a health education program designed to teach rural people how to avoid waterborne disease through the practice of better health habits. This is a policy level decision. It will thus concern the actions of top officials in the central office of the ministry who will establish a new health education unit to coordinate and oversee the work of the program on a national basis. District level officials will then oversee the detailed workings of the program for each of their regions as described in objective two. Objective three describes the need for members of the ministry of health on all levels to work with officials from other ministries. A case in point might be to work with the ministry of natural resources as it employs sanitation experts who possess the technical expertise to combat waterborne parasites that cause the illness in question. Finally, objective four describes how community based personnel will conduct training activities and collect data from rural people to confirm and further analyze training needs for the program.

Once the program goals and objectives are specified, they are further reviewed to determine what personnel will execute these tasks, what the tasks consist of, and if training is required to insure maximum performance. This type of analysis leads to the next design phase, which is the curriculum level.

Curriculum(s) specifies required training courses for a given target audience, such as clinic nurses or by general subject area, such as management development. Clinic nurses, for example, might require training in supervision, recordkeeping, patient care, and counseling. A subject area such as management development might include courses in effective business writing, stress management, and in how to make group presentations. In the latter case, training would not be classified by personnel category. If both staff and managers or individuals from different product lines required similar training in time management, all would attend the same session.

CURRICULUM: Management Development	COURSE: Effective Business Writing
	COURSE: Stress Management
	COURSE: Group Presentations

The following example lists curriculum requirements by target audience for the health care program objectives example.

PROGRAM OBJECTIVE #1.
TARGET POPULATION.
Newly appointed nation health education coordinator

Goals and Objectives

CURRICULUM GOAL
The health education coordinator will be able to plan and implement a national health education program via the creation of a national health education unit.

CURRICULUM OBJECTIVES
The health education coordinator will be able to describe the rationale and expected outcomes of the health education program on a national basis.
The health education coordinator will be able to manage the creation and national operation of a national health education unit.

LIKELY TRAINING AREAS
Program Orientation
Management Development Skills

PROGRAM OBJECTIVE #2.

TARGET POPULATION
District Health Coordinators

CURRICULUM GOAL
District Health Coordinators will be able to plan and implement a district health education unit.

GENERAL OBJECTIVES
District coordinators will describe the rationale and expected outcomes for their district for the health education program.
District coordinators will provide inservice training for clinic health nurses.
District coordinators will manage the day to day implementation of the program in their district.

LIKELY TRAINING AREAS
Program Orientation
Instructor Skills
Management and Supervision

PROGRAM OBJECTIVE #3.

TARGET POPULATION
Newly appointed district level interministerial task force.

CURRICULUM GOAL
District level interministerial task force will be able to act as an intermediary between central and community level ministries.

CURRICULUM OBJECTIVES
Task force members will be able to describe the rationale, expected outcomes on a national and district level and expected involvement for each of their expected agencies.

Task Force members will be able to work together as a group to solve problems and share resources.
LIKELY TRAINING TOPICS
Program Orientation
Group Process Skills

PROGRAM OBJECTIVE #4.
TARGET POPULATION(s)
Clinic Nurses
Rural Health Visitors
CURRICULUM GOAL(S)
1. Clinic Nurses will be able to train and supervise community-based personnel to gather information and conduct training related to the program.
2. Rural Health Visitors will be able to utilize community networks to gather information and to conduct training related to water use contact and sanitation habits.
CURRICULUM OBJECTIVES (NURSES)
Clinic nurses will be able to describe the rationale and expected outcomes for the program for rural people in their service area.
Clinic health nurses will train community based personnel in information gathering and improved practices related to water use and sanitation habits.
CURRICULUM OBJECTIVES (RURAL HEALTH VISITORS)
Rural Health Visitors will be able to describe the rationale and expected outcomes for the program for rural people in the communities they serve.
Rural Health Visitors will be able to gather data, orient community leaders and develop community support for the health program in communities where they serve.
LIKELY TRAINING TOPICS (NURSES)
Program Orientation
Improved practices related to water contact and waterborne diseases.
Instructor Skills
Management Development
LIKELY TRAINING TOPICS (RURAL HEALTH VISITORS)
Program Orientation
Data Collection and Recordkeeping
Improved practices related to water contact and waterborne disease.
Community development and motivation

The above material thus identifies personnel who are assigned to carry out the program objectives that were discussed in the above list. In this example, we will assume that all these individuals require training of some kind to achieve maximum levels of performance in their jobs. Let's look at the first objective, which involves the creation of a national health education unit. We can see that a national education coordinator will be newly appointed who will be responsible for implementing this program objective. If we continue to look at the above items, we can see that this individual's curriculum requirements will involve some kind of training to orient him or her to the need and overall expectations of the total program. In addition, this particular individual requires management and supervision training in how to administer a health education program of this size and magnitude. Decisions and training specifications for the rest of the program objectives follow the same system of analysis.

Let's also look at program objective number four. In this instance, the skills from two different target populations, nurses and rural health visitors, are required to meet this objective. In turn, two different curriculum goals, one for each group, are stated with accompanying objectives.

Once curriculum objectives are identified, they can be used to design individual courses. An instructional analysis occurs on the curriculum level to determine what are the prerequisites necessary to meet each objective. Identified prerequisites can then be used to identify and sequence training topics that will be used as a basis for the development of individual courses. A similar analysis occurs in developing the individual course. Prerequisites necessary to meet course goals are identified, sequenced and converted into learning objectives. A discussion of instructional analyses on both the curriculum and course levels is detailed in a later chapter on sequencing.

At this point, scheduling plans for course design and development are generally established. Priorities identified in the needs assessment phase are reviewed to determine which courses for which curriculum goals and objectives should be developed first. In addition, plans are reviewed to determine who needs to be trained first. A rule of thumb that may guide scheduling plans is to train supervisors before staff. Supervisory support is important to facilitate and ensure the capability of subordinates to do their jobs. Trained supervisors are also likely to be more supportive of the overall program, and thus more willing and able to assist staff in carrying out new responsibilities.

A careful analysis of each target audience should be made next to determine how much is already known about each subject area and what specific concerns within a content area must be addressed for each group. In some instances the training needs may be generic enough to permit the same course to be offered to different target groups. In such instances the design of the course is the same for all groups, even though the actual presentation of the training may or may not occur at the same time. In the above example, orientation training for the national and district health educators and the district level task force may be similar. Let's assume that rural health visitors possess, however, a less sophisticated knowledge of public health care practices and thus require a different orientation approach to the expectations and organization of the new program. This example would require not only different presentation dates, but also the design of two different courses.

It is also important to note that training needs may be technical and non-technical in nature. In the above example, clinic nurses require technical training in new medical practices as well supervision and management training that is less technical in nature. In most cases a combination of the two types of training is required as few jobs are uniquely technical in nature. Most of us require the development of interpersonal and managerial skills to maximize our efforts.

Let's walk through a second example from a different application area. This example concerns the manufacturing of parts for aircraft engines. The company has recently been losing revenue because too many of their parts failed to pass inspection prior to being installed. The discard of so many parts has caused increased overhead costs because of lost personnel time, not to mention the need for additional raw materials. In assessing the problem, findings indicate that both errors and costs could be reduced through better quality controls as well as better trained line employees. While supervisors do possess required technical skills, it was additionally found that they do not adequately supervise

Goals and Objectives

and train their workers. The following lists training goals and objectives on both the program and curriculum levels.

PROGRAM LEVEL

PROGRAM GOAL
To develop an educational program that reduces the incidence of discarded parts or products.

PROGRAM OBJECTIVES
1. To institutionalize the practice and procedures required to conduct quality control inspections that can catch defective equipment before it is completed and ready for installation.
2. To enhance the capability of supervisors to supervise their employees.
3. To improve the technical capability of assembly line employees who manufacture the parts.

CURRICULUM LEVEL

PROGRAM OBJECTIVES #1 and 2.
TARGET POPULATION
Supervisors
CURRICULUM GOAL
Supervisors will be able to contribute to final product quality through better training, supervision and evaluation of workers and their products.
CURRICULUM OBJECTIVES
Supervisors will be able to conduct better and more timely quality control reviews.
Supervisors will be able to conduct on-the-job training for workers on required technical skills.
Supervisors will be able to better supervise and manage employee work.
LIKELY TRAINING AREAS
Quality control planning and testing
Mentoring skills
Management and supervision

PROGRAM OBJECTIVE #3.
TARGET POPULATION
Assembly line workers
CURRICULUM GOAL
Workers will produce fewer deficient products.

CURRICULUM OBJECTIVE
Workers will employ required technical skills to correctly complete their work.

LIKELY TRAINING AREAS

Technical Skills

In the above example the real training focus is on the supervisors. While the immediate cause of the problem lies with the production of deficient parts, the real roots appear to lie with the supervision and evaluation of the work effort. It thus follows that workers need proper direction and support to produce the desired level of product. Similar situations are typical. It is not uncommon to find that workers cannot properly do their work because they are not properly managed.

Micro Design

Micro design is the more detailed type of analysis required to develop individual courses and lessons. The design of courses and lessons is closely linked with performance requirements that individuals must fulfill to successfully carry out their designated role in the overall program. Goals and objectives are thus derived from performance expectations and are written as observable and measurable behaviors. In turn, these behaviors are written in terms of what the "learner" will be able to do after successfully completing the instruction. The word performance is important because goals and objectives state what behavior or performance the trainee must exhibit (write a paragraph, draw a triangle) to signify that he or she has mastered the material.

This approach differs from traditional instruction, in which plans are generally driven by what the instructor or student will do during instruction. Thus the statement, "The teacher will show a ten minute videotape on procedures for operating an electric drill" is not a proper statement of a lesson objective because it refers to what the TEACHER will do DURING the lesson. In turn, the next statement is equally inappropriate, "The student will view a ten minute videotape on procedures for operating an electric drill." In this statement, the reference is in terms of what the STUDENT will do DURING the lesson. Both statements do not address the question of WHAT the student will be able to do AFTER instruction, but rather describes HOW the learning is to take place. It would be acceptable to have an objective if it were worded, "The student will demonstrate how to use an electric drill to make holes of various sizes in wood." This statement is more useful because it is learner-centered and describes actual performance expectations resulting from the instruction.

In designing courses, the first step is to develop clearly worded goals. Goals differ from objectives in that they describe the major culminating or synthesizing behavior which results from completing the course. Often initial attempts at identifying instructional goals result in goal statements that are vague and loosely worded. Such statements describe the desired outcomes of instruction in nebulous terms that are open to much interpretation and confusion. For example, the statement, "The student will appreciate the democratic form of government" could mean many things to different people. How would we know if the goal was achieved? Must the students define a democratic form of government, describe why it is important to vote, compare and contrast democratic and authoritarian forms of government, identify major features of a democracy, choose to vote in an election, become better informed on key issues, or identify countries with democratic governments? The situation also applies to such vague statements as, "The student will develop a positive attitude and respect for the opinion of others" and "The student will appreciate good music." It is thus necessary to develop more specific goal statements from these vague or "fuzzy" statements.

A procedure termed goal analysis has been developed to translate these "fuzzy" goals into more tangible, specific goals. Mager (1972) suggests five steps to help clarify goals. Briefly stated these are:

1. Write down the goal.
2. Write down the performances that define the goal.
3. Go back and tidy up the list, crossing out duplications and items that are not what you want to say.
4. Describe each performance by identifying the manner or extent (or both) to which it must be carried out in order to show that the goal has been achieved.

5. Modify each resulting statement until you can answer "yes" to this question: "If someone fulfilled the requirements of this performance, would I be willing to say he had achieved the goal?"

Let's walk through the process by returning to the example of a vague statement that we gave before:
   1. The goal.
      "The participant will develop a positive attitude and respect for the opinion of others."
   2. Performances that define the goal.
      a. Listen to someone else without interrupting
      b. Correctly repeat the ideas and opinions of someone else.
      c. Don't talk while others are speaking.
      d. Name other people who agree with this individual.
      e. List good deeds that this person has done.
      f. Describe the positive aspects of another person's opinion.
   3. Cross out duplications and items that do not apply.
      a. Applies
      b. Does not apply
      c. Duplicates
      d. Does not apply
      e. Does not apply
      f. Applies
   4. Describe what must be done to show that the goal has been achieved.
      a. Listen to someone else for five minutes.
      b. Explain how this set of ideas can benefit something or someone.
   5. Describe the positive aspects of another person's discussion by listing how these aspects can benefit something or someone else. Listen without interrupting for five minutes so that the other person can explain his or her ideas.

Goal analysis thus helps to clarify what it meant by success. It also represents one of the greatest investments of time and creativity of any of the initial steps in course design.

Once course goals have been determined and clarified, they are used as a basis for identifying required content and sequence. Until fairly recently, decisions about course content were made by subject matter experts (SMEs) in the particular field. An in-

Goals and Objectives

structor would decide what topics to cover and how to sequence them. He would then select a textbook that included these topics. In most instances textbook authors and instructors try to select and sequence the content of their courses so that students first acquire some of the simple ideas and then build towards the more complex issues. However, this is seldom systematic. When faced with the task of specifying content for the same instructional goal, 20 instructors or subject matter experts may identify vastly different content and instructional sequences. In turn, the risk of leaving out crucial content is higher.

Beginning in the mid-1950s, attention and research began to focus on identifying the precise skills and knowledge which should be included in courses for students to effectively and efficiently reach specified goals. This led to the concept of prerequisite skills or knowledge.

Prerequisites are of two sorts: *essential* and *supportive*. Essential prerequisites are those which must be learned before the learner can progress to the next capability. For example, a student cannot MULTIPLY multi-place numbers without first possessing the more elementary capability of ADDING multi-place numbers. Supportive prerequisites are those which, while not absolutely necessary, may aid the new learning by making it easier and faster. For example, a positive attitude towards communicating precise meanings to others may aid in the learning of sentence construction.

Several procedures have been developed for identifying these prerequisites and are discussed in Chapter 9. These procedures constitute a part of course design called instructional analysis and permit the designer to systematically identify WHAT to teach and in what SEQUENCE to teach it.

Once prerequisites have been identified and sequenced, they are converted into learning objectives. As stated earlier, objectives should be written in behavioral, observable and measurable terms. Mager states that an objective is "a description of a pattern of behavior or performance that we want the learner to be able to demonstrate." He stresses three main components:

1. Performance—what the learner will be able to do.
2. Conditions—important conditions under which the performance is expected to occur.

3. Criterion—the quality or level of performance that will be considered acceptable.

PERFORMANCE. Objectives must describe what the learner will be expected to do. This is the only way a skill can be assessed. One cannot look inside a person's head to see what he or she knows. To say that the trainee or student "knows how to read German" or that he or she "understands plane geometry" creates a great deal of ambiguity. Words like "understands" or "knows" are open to a wide variety of interpretations, whereas words like "list" or "write" or "state orally" are observable and measurable and thus are much clearer.

CONDITIONS. It is helpful to specify the conditions that will be imposed upon the learner when he or she carries out the specified performance. For example, if the learner is to "be able to obtain the sum of five single-digit numbers," would you have the trainee write on a piece of paper, punch keys on a calculator, or count on his or her toes? This statement provides the learner with clear, unambiguous expectations, and is usually referred to as the "given" part. For example:

"Given the use of notes and references . . . ."

"Given a hammer and a saw . . . ."

Conditions may also specify limitations imposed upon the learner. For example:

"Without the aid of an electronic calculator . . ."

Note that the conditions of an objective do not refer to the conditions under which the learning occurs ("after listening to a lecture") but to the conditions surrounding the performance.

CRITERION. This component is the performance standard which is used to decide whether the learner has mastered the objective. Note the following sentence fragments:

- ". . . must be able to solve three quadratic equations within ONE HOUR."
- ". . . . . . . must be able to ride a bicycle ONE MILE IN SIX MINUTES."
- Here is a complete objective written in Mager's format:
 "Given 10 linear equations to solve (and no outside references), the student will write both the steps in the solutions as well as the answers, getting eight of ten answers

Goals and Objectives 109

correct, and without missing any steps in the solution."

Guidelines for Developing Goals and Objectives

Functions and Benefits

Taking the time to clearly identify goals and objectives on the four different levels is thus important as they play an essential role in the design process. They are used to identify training outcomes, guide content development, monitor participant progress and aid in course evaluation.

The specification of training outcomes assists both the training participants and the organization's management. First, participants know what is expected of them and thus are better able to focus their efforts. Knowing what is expected can minimize failure and frustration because less time and energy are wasted on extraneous material. It also permits a less subjective type of learner assessment. By stating the requirements for mastery in advance, the student knows what has to be done and then chooses to do it. In turn, this process permits a uniformly high level of achievement among all participants, since learners are not ranked on a bell shaped curve as in traditional instruction. Instead, they are assessed on their *individual ability* to master goals and objectives as specified.

Participant progress can also be better monitored. Instructors can assess how far the learner is from the desired outcome, what areas are giving them the most difficulty, and determine if they are progressing at a proper pace. As a result, the instructor can better identify what parts of the instruction require remediation.

Decision-makers are also given a clearer idea of what they are getting for their investment. By having a more tangible understanding of what the program is expected to achieve; they can voice their satisfaction or dissatisfaction early on. As a result, revisions can be made while the training is still being planned, thus saving the expense of more costly changes later.

The specification of goals and objectives is also used to guide the initial choice and sequencing of content. As discussed earlier the systematic analysis of terminal outcomes and their prerequisites better ensures inclusion of required content in the right order. This can contribute to substantial cost and time savings due to elimination of unnecessary content. They also provide a sound basis for the subsequent selection of instructional methods and media as discussed in Chapter 10.

Finally, goals and objectives can aid in evaluating all four levels of the design process. They can be used initially as evaluation criteria to assess needed revision of instructional plans and materials. Later, they can be used to gather follow-up data in evaluating the overall success of the program to meet ultimate outcomes.

Selecting the Right Goals and Objectives to Pursue

Not all training needs are worth developing. To determine which ones are worth spending the time, expense and energy it takes to develop materials and conduct training, it is helpful to examine the following questions:

1. How stable is the content?
2. How large is the target population?
3. Are important behaviors stated?
4. Can the goal be reached in a reasonable amount of time?

One of the first concerns that should be addressed is the stability of the content. This is especially true if the subject matter is likely to change in a short amount of time. For example, why train people on how to use an electric typewriter if they will be doing computer-based word processing six months from now? Virtually all training will, of course, need to be updated from time to time because of changes in policy, rules and regulations, and the technology itself. This issue is related to course maintenance and is different from the question of whether to design the material in the first place. When possible, it is best to design training so that only parts of it need to be revised at any one time. In most instances, courses can be modularized to permit sections to be independently changed without disturbing the rest of the instructional package.

Another consideration is the size of the target audience. In general, the more people need the material, the more sense it makes to develop it. The exception, of course, is related to the criticality of the training. If enabling a certain group of people to carry out a certain role is essential to meet program level goals,

Goals and Objectives 111

then the number of people involved might not matter. For example, a group of information systems managers may require cost-benefit analysis training to be able to make better decisions about the purchase and installation of new hardware and software. As cost-benefit analysis and planning can affect the ultimate profitability of their work, it is likely that this kind of training will be considered crucial, even though the number of managers involved is small.

Two more issues are important once the designer reaches the level of micro design. First, goals and objectives need to be stated to ensure that they represent important behaviors. For example, a designer might choose to identify a terminal behavior in terms of problem-solving capabilities rather than memory skills. Thus, the listing of principles for good dietary planning may be less important than the ability to develop a customized diet for a hospitalized patient. The rationale, of course, is that the ability to use information to problem-solve is a more important and advanced skill than memorization.

Second, goals need to be examined to determine if they can be reached in a reasonable amount of time. Dick and Carey (1985) suggest that material should be organized into instructional chunks whose goals require no more than 15 hours to complete. This is not a hard-and-fast rule as many courses will last for a week or two at a time. However, a tendency to provide training in shorter doses can be both helpful and practical. In most training situations, participants can be released from their duties to attend classes for only a limited amount of time. Organizing material into smaller units permits additional ease in course revisions. Also, shorter sessions permit the opportunity to practice and receive feedback on smaller amounts of content both in class and once they return to their jobs. Finally, learning new skills and practices in smaller doses tends to make the training more effective and less tedious as participants are required to process and remember less at a time.

Research Data on Goals and Objectives

The following data summarize major research efforts on presenting goals and objectives to learners prior to instruction. We

have additionally tried to cite some of the more recent studies that support these findings.
1. They will clearly reduce study time.
2. Studies are either pro or null on whether they will increase learning. (Duchastel and Merrill, 1973)
3. They can limit incidental learning; but will strengthen intentional learning. (Merton, 1978)
4. They tend to benefit lower level learners more.

Research data indicate that communicating objectives to learners before the instruction begins will clearly reduce study time. This finding is associated with the learners' increased ability to focus their efforts. Time is less likely to be lost studying extraneous material, which in turn can lead to less frustration. Reports, however, are conflicting as to whether the use of objectives will increase the amount of learning. However, there is no evidence indicating a decrease in learning! Thus, studies are either pro or null in favor of goals and objectives.

Studies also indicate that the use of objectives can limit incidental learning but will strengthen intentional learning. Intentional refers to learning that is directly related to the learning objectives. Incidental learning, on the other hand, is unplanned and tends to occur when lesson plans are less structured. An example of incidental learning might be associated with the case of an international businessman whose objectives are to describe the geography and products of a given country. However, the instructional materials might also include information on the country's climate and topography. For example, a given product may be produced because it is warm, humid, and mountainous in part X of country Y. If the learner focuses only on the objectives, it is probable that he will not learn why the product grows in that part of the country.

A final research finding is that objectives tend to make the most difference with students who are less bright. The theory is that bright students require less assistance and direction and are able to create organization where none may exist. Less able students, however, appear to need more structure and guidance in order to maximize their study efforts.

Goals and Objectives 113

Summary

This chapter described the use and development of goals and objectives to specify training outcomes. This part of educational planning is called the design phase. Its role is to translate performance gaps identified in the front-end analysis into goals and objectives to be accomplished by the learner as a result of the training. These expectations can be stated in both ultimate and process terms. Ultimate outcomes refer to the final mission of the training, whereas process outcomes refer to the coping capability of individuals and organizations to achieve expected results.

The design process was described as occurring on four different levels. They are program, curriculum, course and the individual lesson. Program and curriculum levels are linked with macro or long-range planning. Micro design is associated with the different kind of analysis and planning that occurs in the design of courses and individual lessons.

Program goals and objectives describe how training can be implemented to help reach societal and organizational needs such as a stronger national defense. Training plans often constitute only one component of a larger effort to reach the same outcome. Because they are generally linked to ultimate outcomes, it may take several years to complete this kind of long-term development. Curriculum planning, then, converts training needs into courses for personnel designated to complete certain tasks required to achieve program level goals.

Course level design requires careful analysis of goal and objectives to ensure that they are written as observable and measurable behaviors. Mager's strategy was discussed as a means to help ensure that goals are written in non-fuzzy terms. Once goals are specified, an instructional analysis is completed to identify prerequisite skills that can be converted into learning objectives. Lesson plans are next developed to implement these objectives.

Guidelines for developing goals and objectives on all four levels were next reviewed. They can be used beneficially to identify training outcomes for both learners and decision-makers, guide content development, monitor participant progress and aid in course development. Not all goals and objectives, however, are worth pursuing. The following factors must be reviewed in the selection process: stability of content, size of target population, importance of stated behaviors, and the amount of time it takes to reach identified goals. Finally, research data were discussed as applied to the use of goals and objectives.

Chapter 8
Analysis of Intended Participants

Since the purpose of any educational/training program is to teach some content to some persons, it is important that we carefully consider both *what* we are going to teach and *who* we are going to teach; the analysis of the intended learners defines who we are going to teach. The purpose of this chapter is to explore various techniques for getting and using information about the intended learners in the development of educational/training programs.

Prior to the development of courses it is essential that we get a good "fix" on the learners for whom the education/training programs are intended. If we know about the learners in advance, then we can develop education/training programs that are understandable and meaningful to the learners. That is, is the instruction too hard? Too easy? Are the examples appropriate for this group? Will they understand? When this knowledge about the intended learners is used in the actual education/training program development, these programs will most likely be more successful. When you develop instruction in such a fashion you are writing for the specific audience rather than writing to the content. All the instruction is designed to be understood by the audience for whom the instruction is intended. While the lessons that comprise a training program must have content that is accurate and covered well, this is not sufficient. The lessons must be understandable to the learners; they must communicate effectively.

Rationale

Before examining the different kinds of information about the intended learners, let's look at the rationale for considering the intended learners before developing the education/training programs. The approach often followed in developing instructional materials is to "write to the content." Often when developing instructional presentations, the authors, who are usually specialists in the content, fail to consider the intended learners and focus instead on the content itself. Perhaps they consider the subject matter as a self-contained, stand-alone entity. This may account for the failure of some instruction; it was not understandable to the learners. The person presenting the instruction may have covered the content but not in a form understandable to the learner. The recommended approach is to write to the audience, the intended learners, not to the content. In order to do this you must "know" your audience in advance.

Role of prior knowledge. There is evidence to suggest that the most important factor determining whether instruction will be successful is what the learner already knows (Ausubel *et al.*, 1978). The instruction will be meaningful when the learner possesses some information relevant to the topic of the instruction. When the learners have some knowledge to which the new knowledge can be related, there is a greater probability of learning and retaining the new knowledge.

Determining where to begin instruction. Once we have specified the instructional objectives for our courses, we must determine at what level to begin the instruction. For example, if the topic is appropriate use of adverbs, do we include instruction on verbs and adjectives? Do we have to provide instruction on what a sentence is? Certainly we would proceed differently in the development, depending on what the learners already knew about these topics. If we conducted a careful analysis of our intended audience, we would have information about what the learners already knew and could use this as a basis for deciding where to begin the development of our lesson.

Determining how to present instruction. Information about the learners can also be helpful in determining how to present new

Analysis of Intended Participants

instruction. For example, we might develop different approaches to teaching first aid if one group of learners were college educated nurses and another group were poorly educated youth working on an assembly line. Not only would the starting point differ for these groups, since their knowledge of first aid differs, but since the groups also differ in degree of literacy, the instructional presentations would be varied accordingly. The lesson for poorly educated persons would have to rely on a more visual presentation and use a very restricted vocabulary. Certainly the instruction would have to be more concrete than the instruction developed for nurses. Since these groups differ both in terms of what they know relative to the lesson topic and in terms of their abilities to learn, the lessons for each group should differ in terms of what is presented and how it is presented.

In summary, information about the intended learners can help us in developing lessons in two primary ways: (1) by assisting us in determining *where to begin* the content presentation, and (2) by assisting us in determining *how to present* the content so that the learners can understand the presentation. Since this knowledge of the intended learners is so crucial to the development of education/training programs, it is important to consider what kinds of information should be collected about the intended learners and how this information might be collected. The remainder of this chapter considers these two points.

Entering Characteristics

The knowledge and skills that the learners bring to the instructional situation are called their *entering characteristics*. The assessment of entering characteristics is a crucial step in developing any instructional materials. The first step in assessing entering characteristics is to identify the specific characteristics that may influence the development of the instruction. There are three different categories of entering characteristics that can be used in developing education/training programs. These are:

(1) general characteristics of the learners
(2) specific entry behaviors
(3) learning styles

Each of these will be presented in this chapter.

General Characteristics

Certain general characteristics of the learners can influence the way that the education/training programs should be developed. The way you would instruct a group of teenagers would likely be different from the way you would instruct a group of retired persons, even though the topic may be the same. The contents of an education/training program on accounting procedures would be quite different for each of the following groups: small business owners with college degrees, CPAs, high school students, and a group of adults with no college education.

Age and prior educational attainment of the learners are two of the general characteristics that the developers must know prior to creating the education/training programs. The interests that the intended learners have can also influence the development of the education/training program. Likewise we would develop the education/training programs differently for a group of learners with a high level of intelligence than for a group with normal or lower levels of intelligence. We could develop education/training programs that progress more quickly for the gifted and talented learners. For learners with lower levels of educational attainment the education/training programs would progress more slowly and include more prompts and cues in the instruction to assist the learners.

Specific Entry Behaviors

Another type of entering characteristic that is helpful to know about in planning instruction is the learners' specific knowledge and skill related to the objectives of the education and training program. These knowledge and skills are referred to as *specific entry behaviors*. These are the behaviors related to the objectives that the learner has already mastered. The specific entry behavior is specific to the objectives of a given course rather than some general level of achievement. Thus the ability to read at a fifth grade level is a general characteristic; the ability to distinguish plural from singular nouns is a specific entry behavior.

Learning Styles

The final type of entering characteristic that we might consider in developing education/training programs are the *learning*

Analysis of Intended Participants

styles of our learners. Are they capable of abstract, symbolic thought? If not, then we would have to develop a presentation that is more concrete, making greater use of visual images. If the learners are independent and have well developed abilities to structure their study, we may create a flexible type of education/training program that gives them choices about how and what they study in the lesson. For less independent learners, we would develop a more structured education/training program that gave them few choices about how to proceed. Some learners may prefer a visual presentation while others prefer a verbal presentation.

A number of different learning styles or cognitive styles that may have implications for training have been identified (Messick, 1984). These styles are based on how individual learners approach a learning task or how they process information. Some speculate that one type of person may learn in a predominantly visual fashion, while another may learn in a verbal fashion. Another style that has been postulated is whether the learner operates in an inductive or deductive manner when approaching a learning task. Does she go from specifics to wholes (inductive style) or from the whole to the specifics (deductive style). Another style is whether a person focuses on the differences among stimuli (a sharpener) or the similarities among stimuli (a leveler). There are numerous examples of other learning styles and cognitive styles. On the surface each style would seem to imply a different instructional approach, e.g., visual learners should be given visual, not verbal, lessons. The intent in using information about the learning style or cognitive style of a trainee is to somehow adapt or adjust the training to take advantage of his or her unique style.

Research evidence. Unfortunately, the research evidence on learning styles is quite mixed. For all its intuitive appeal, it is rare to find clear examples of these styles that significantly influence the ability of a person to learn when his/her style is not attended to. A similar situation exists with learner preferences. While learners do show preferences for different instructional approaches, these preferences seem to have little if any relationship to the amount of learning that occurs (Clark, 1982). Thus, a word of caution must be added to this discussion of learning styles. There is not clear evidence that learning styles are all that important in determining whether someone will be successful in learning. While it seems essential to adjust the instruction depending upon the general characteristics and specific entry behaviors of the learners, the use of learning styles as the basis for developing lessons is questioned.

Assessing Entering Characteristics

There are several ways to identify or assess the various entering characteristics of the intended learners. Practical constraints usually dictate that we can't assess the characteristics of each person from the intended audience. Fortunately, such total assessment is not essential. Rather, the usual situation is to select a representative sample of learners from the intended audience and assess their entering characteristics. Such is the approach recommended here. Don't attempt to gather information from each potential learner, but do get information about the entering characteristics from a sample of the learners. This is like a political poll in which a few voters are asked their preference among candidates for office and the results generalized to the population as a whole. As we have seen, carefully constructed polls can be very accurate in predicting the choices of all voters. Likewise, we can use a sample of learners to get an estimate of the entering characteristics of the population of intended learners. We will be accurate in the estimate to the extent that the sample represents the population of intended learners. You must exercise care in identifying the learners to be included in the sample. The best approach is to select from all learners at random; thus each learner has equal probability of being included in the sample. If a random selection is not possible, then be careful that the sample represents the whole group. Try not to select just the "better" or "worse" learners.

Assessing general characteristics. Gathering information about the general characteristics of the intended learners is usually rather easy to do. Descriptions usually exist of the demographic characteristics of the learners. Thus, you can find out about the ages, sex, prior educational attainments, etc., of the learners. Descriptive information about the intended learners can be ob-

Analysis of Intended Participants 121

tained through records or from conversations with persons familiar with the learners. If necessary, it is acceptable to use information about past learners—about last year's class—as a proxy for information about the next class of learners. While there are some changes from year to year or group to group, the general characteristics of former learners approximate the characteristics of current learners. This information is not perfect but can be more helpful than not having any information.

Assessing specific entry behaviors. Information on the specific entry behaviors of the intended learners is not usually available. Therefore, you probably will have to administer tests of the specific entry behaviors to a sample of intended learners. In the process of defining the objectives and analyzing the content for your education/training program, you would have identified the specific knowledge and skills that learners should possess to benefit from the education/training program.

Let us return to the example about the ability to distinguish plural from singular nouns. This ability is an essential prerequisite ability to learning how to use verbs or pronouns that must correspond to the noun in number. Therefore, if the intended learners have this ability as an entry behavior, the education/training program can proceed to build upon it by starting with the concept of the verb agreeing in number with the noun. However, if the intended learners do not have this ability prior to beginning the instruction, then this must be taught first in the education/training program. Thus, knowledge of the intended learners' specific entry behaviors is crucial in determining where to begin the education/training program, determining what content to include, and what content to assume the learners have had.

While it may be tempting to talk to instructors that the intended learners have previously had, review course descriptions, or examine the books they used, all of these approaches to estimating specific entry behaviors are much too risky and should be avoided. There seems to be only one acceptable way to estimate entry behavior: test specifically for it! Perhaps the learners didn't read all of the textbook that was used, or they didn't retain all they had read. Thus, the textbook would have given an inflated estimate of the learners' specific entry behaviors. While we like to assume that what we present as instructors is learned and remembered in its entirety by each learner, such is rarely the case. Consulting the instructor about what specific entry behaviors we can expect from the intended learners will yield an inflated estimate also. "Oh yes, we covered that in class, they will be able to do that." These are common words from prior instructors. Unfortunately, there is much danger in assuming that the intended learners are, in fact, capable of such entry behaviors. Test a *representative* sample of learners to determine what specific entry behaviors they do possess. There is no acceptable substitute for test data.

Assessing learning styles. There are two ways of determining the learning styles of the intended learners. For some learning styles it is adequate to simply ask the learners what they prefer or which style they have. Of course, to be useful in developing education/training programs you would have to ask a representative sample of learners so that you could make a judgment about the group as a whole and then use this information as basis for education/training program development. The second way of determining learning style is to administer any of several learning style inventories to a sample of intended learners. There are quite a few inventories or tests of different learning styles that could provide information to persons developing education/training programs who wish to alter their education/training programs to match certain learning styles.

Summary

What the learners bring with them to the instructional situation is an important determinant of the effectiveness of the education/training program. We should consider both the general characteristics of the learners and their specific entry behaviors as the education/training programs are being developed. While the evidence is not as clear-cut, you may use information about learning styles as a basis for designing the approach used in the education/training program.

Chapter 9

Organizing and Sequencing Instructional Content

This chapter will present an approach to organizing and sequencing instructional content based upon research on learning. There have been many studies about how people learn that can serve to guide how we should organize and sequence instruction. Since the intent of this book is to focus on designing large-scale educational/training programs rather than designing specific lessons or instructional units, the closest we will get to lesson design is this chapter on organizing instructional content. Although this book is written primarily for persons with responsibility for planning educational/training programs rather than for developing instructional materials per se, it is important to be aware of the kinds of training outcomes that can be sought and the ways of organizing the instructional content to achieve these outcomes. The overall focus of this book remains at the program level and on the "big picture."

Organizing Instructional Content

The starting point in organizing and sequencing instructional content for education and training programs is to assign large units of related content to courses. Then the related content within each of the courses is grouped into several individual lessons. Finally the content of each lesson is analyzed to determine the necessary supporting content or prerequisites. Once the content of individual lessons has been determined, the instructional events that are the component parts of a lesson are generated and sequenced. Figure 9.1 shows the relationships among programs, curricula, courses, and lessons.

Figure 9.1 Levels of Programs, Curricula, Courses and Lessons.

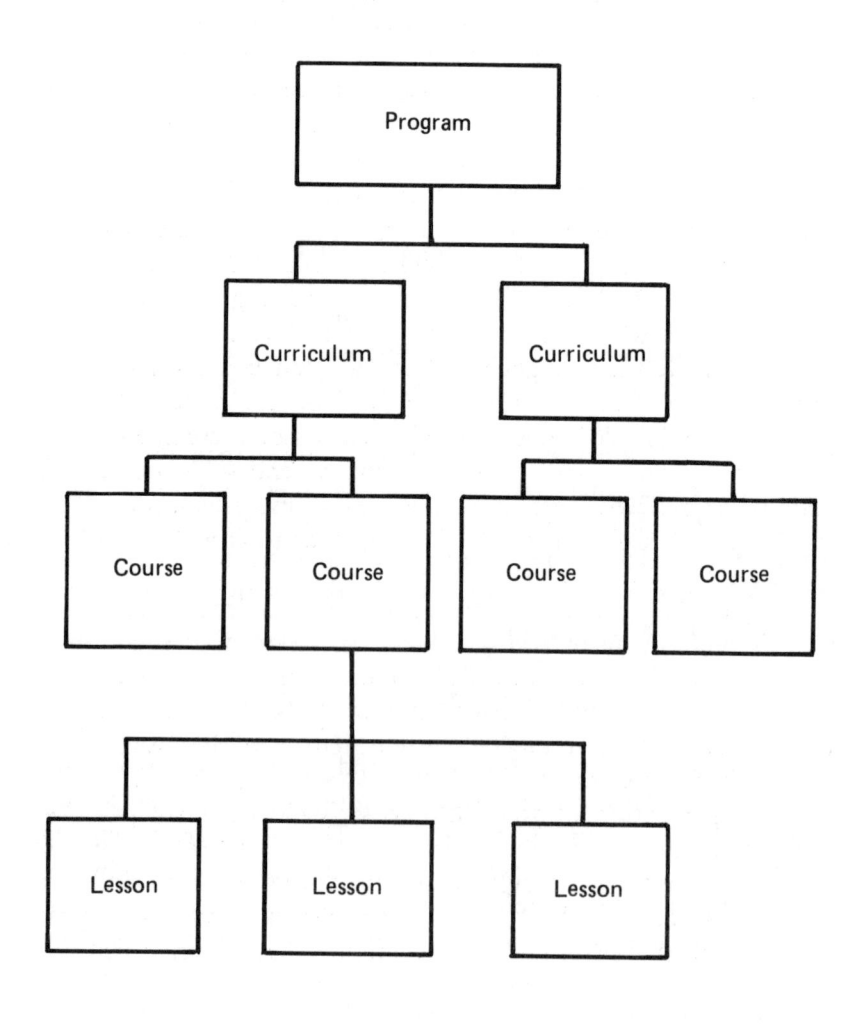

Courses. As you begin to organize and sequence the courses for a given curriculum, your attention first focuses on the expected outcomes stated as training or instructional objectives. The first step is to organize related goals into course sized groups. In different education and training programs, courses may consist of a different amount of instruction. For example, some curricula may last for three months and have several week-long courses. Other education and training programs may last for two years and have courses that run for several weeks. Of course, the number of hours spent in instruction may be the same in each course. In a typical university course students attend three hours of class for 14 weeks for a total of 42 instructional hours. The educational curriculum of the university is made up of many such courses. An industrial training curriculum could be composed of courses of an equal length, 42 hours, but a person in training may devote full time to a course, completing one course a week. There is some variance in the time assigned to curricula and courses in different settings. In general a lesson usually can be completed in 1 to 3 hours; a course contains several such lessons completed over several days; and a curriculum contains several such courses that may be completed over several weeks if intensively done, but more often curricula require several months to complete. In some instances, they will span several years.

Lessons. The second step in organizing instructional content is to examine the instructional objectives assigned to each course and arrange related objectives into individual lessons. Each lesson should contain a few objectives that are similar. The content in a lesson must be internally consistent. Often each lesson would treat only a single concept.

The organization of goals into courses and lessons is primarily a judgment made by content experts. This is a logical grouping. Once the objectives are organized into lessons there are analytical techniques that can be used to help identify and organize the content for the lessons. These techniques have been called task analysis, learning task analysis, and instructional analysis by different authors. The intent is to "flesh-out" the instructional content to support the objectives.

Learning task analysis. The third step in organizing and sequencing instructional content is to conduct such a learning task analysis of the instructional objectives. There are several substeps necessary for this analysis that can be used to take much of the "guess work" out of defining and sequencing the instructional content for given objectives. If the instruction is to be judged successful, then it must focus on teaching the proper content to support the learners' attainment of the instructional objectives. Thus, rather than rushing in and deciding how to teach some instructional goal or what the learner should do during the instruction, time should first be spent in determining exactly *what* the learner must acquire in order to reach the goal. This is where the analysis of the instructional content will help in developing lessons.

The starting point in learning task analysis is to identify what type of learning is implied by the instructional goal. Gagne (1977) has indicated that learning outcomes can be placed into one of the five *domains of learning*. These domains are: (1) information, (2) intellectual skills, (3) motor skills, (4) attitudes, and (5) cognitive strategies. Since the learning outcomes are different in each domain, the analysis of these learning outcomes requires different techniques. Once the particular domain is identified, then the appropriate analysis technique for that domain can be applied (Hannum, 1988).

Information domain. An *elaboration analysis* is the technique recommended for analyzing the instructional content in the domain of information. In this approach the key idea or real essence of the information is first identified. Regardless of how complex and grandiose the information may at first seem, there is some key aspect or central idea to it. This has been called the epitome by Reigeluth (1983). In analyzing information you must first identify this epitome. For example, in a lesson on American government the epitome may be that our government is a representative form of government composed of three branches. Once this core has been identified, then you proceed to identify the next level of information, the information that *elaborates* on the epitome. In the example of American government this elaboration might include a description of how representatives are selected and the functions of the three branches of government. In turn you

Organizing and Sequencing Instructional Content 127

would develop *further elaborations* of this information including more detail each time. In the elaboration approach you first identify the main ideas or the big picture then successively identify the secondary or supporting ideas. First the forest, then the trees. This approach is recommended to: (1) make sure that all the important information is included in the lesson, (2) that no unrelated information is included, and (3) provide the much needed structure of the information for learners. When we can place order or structure on information we tend to learn it better and retain it longer.

Intellectual skill domain. the development of a *learning hierarchy* is the approach recommended for the analysis of instructional content in the domain of intellectual skills. Since the learning of any intellectual skill depends upon the prior mastery of prerequisite skills, then these prerequisite skills must be identified. The approach suggested by Gagne (1977, 1985), is to begin with the instructional goal and ask "What must the learner be able to do in order to reach this goal?" The goal is broken down in stages to identify the prerequisite skills. For example, suppose we wanted to develop a lesson to teach the rule for finding the size of the hypotenuse of a right triangle when given the size of the two legs. Before a learner could so this she would have to be able to sum the squares of the legs. Before she could do this she would have to identify the legs of a triangle and be able to square numbers. Before she could do this she would have to be able to multiply two numbers, etc. In such a manner we have worked backwards starting with the instructional goal and identified the prerequisite skills necessary for accomplishing that goal. These prerequisite skills then are the basis for our instructional content. In our lesson we would include instruction on each of the prerequisite skills and *not* include instruction on other, non-essential topics.

Motor skill domain. A *procedural analysis* is the recommended technique for analyzing instructional goals in the domain of motor skills. Any motor skill is composed of a collection of individual movements and the routine for sequencing these movements. In a procedural analysis, motor skills are broken down into a series of specific steps which when executed comprise the motor skill. The motor skill of starting a car can be broken down into several steps that must be executed in a sequence. For example, place the key in the ignition, make sure the parking brake is set, place the transmission in neutral, depress the gas peddle slightly, turn the key clockwise, and then release the key upon starting. If the instructional goals of the lesson are motor skills, then carefully identify the sequence of steps that must be performed to accomplish this motor skill.

Attitude domain. *Goal analysis* is the recommended technique for analyzing instruction in the domain of attitudes. Since attitudes cannot be directly observed but must be inferred from behaviors, the analysis of attitudes involves the identification of these behaviors. In doing a goal analysis of an attitude, you ask "What is the behavior that would mean that a person possessed this attitude?" If we were developing a lesson to teach a positive attitude towards safety in the workplace, then what behavior would we accept as evidence that a person had the desired attitude? Alternatively, we could think of an employee that had a good attitude towards safety and one who didn't. We could then think of what it was that they did differently to make us think one did and one didn't have the desired attitude. In any case the intent is the same, to describe the behavior that indicates or represents the attitude. In this example, if an employee could verbally state all safety regulations, used safety precautions in his work, could describe how to handle potentially dangerous situations, or could respond in simulated emergency situations, then we may say he had a positive attitude towards safety. These specific behaviors that correspond to the attitude would be the basis for the instruction.

In summary, different techniques must be used to analyze the instructional objectives to identify necessary instructional content. If the instructional goal is in the domain of information use an *elaboration approach*. If the goal is in the domain of intellectual skills develop a *learning hierarchy* to identify the necessary prerequisite skills. If the goal is in the domain of motor skills use a *procedural analysis* to identify each step in the motor skill and the sequence of these steps. Finally, if the goal is in the attitude domain use *goal analysis* to identify the behavior that is indicative of the attitude. Regardless of which procedure is used the purpose is the same—to identify the necessary content that when learned will lead to reaching the instructional goals.

Sequencing Lessons

There are two aspects to sequencing lessons: (1) the sequencing of the instructional content, and (2) the sequencing of instructional events. By instructional content we mean those facts, ideas, concepts, skills, etc., that are reflected in the instructional objectives. This is the content that we expect the learners to master by going through the lesson. By instructional events we mean those features of a lesson that, when present, facilitate learning. These include such things as informing the learner of the objectives of the lesson, providing examples, and use of feedback. A more complete description of these instructional events follows in this chapter.

Sequencing the instructional content. The content in different domains of learning has a different natural organization to it. Motor skills have procedural structures; intellectual skills have learning prerequisite structures. These different structures imply different instructional sequences. Recommendations for instructional sequences are grouped by domain of learning outcome.

Information domain. There are some general "rules of thumb" to be followed in sequencing content in the domain of information. During the days that programmed instruction reigned supreme, all sequencing of instruction was linear from simple to complex. More recently, research in human learning has suggested that this might not be the best way to sequence the learning of information. In fact, almost an opposite approach is recommended. This approach is to first present the "big picture" or main idea then go back and fill in the specific knowledge. In some sense this could be considered a general to specific sequence.

The starting point in the sequencing of information is with the presentation of the central idea or essence of the content. Reigeluth (1983) has called this, as noted above, the epitome and recommends this as the appropriate starting point for instruction. After this epitome is presented, an elaboration of the epitome is made. This elaboration includes additional information that expands upon, or elaborates, the central idea. The information in this elaboration is, in turn, expanded upon further.

In designing such instructional sequences, you should keep referring to the structure of the content so that the learner never gets "lost." A useful way to do this is to move from the central idea to an elaboration of it then back to the central idea, sort of "zooming" out, then back in. Using such an approach ensures that the learner attends to the specifics in the instructional content as parts of a whole rather than in isolation. This, in turn, fosters learn learning and retention.

Intellectual skills. The sequencing of content within the domain of intellectual skills is rather straightforward. The knowledge structures in this domain are learning prerequisite structures that depict the content and inter-relationships in the form of a learning hierarchy. The prerequisite skills are shown at the lower levels of the hierarchy. Since the mastery of the higher level skills is contingent upon the prior mastery of the lower level, or prerequisite, skills, the instructional sequence must begin with the lower level skills and proceed upwards. By proceeding in such a manner, the instruction is likely to be much more effective.

Motor skills. The type of knowledge structures found in the domain of motor skills are procedural structures. These structures show the individual part skills that make up a complete motor skill and they show the sequence in which these skills should be executed. In learning a motor skill a person must learn each discrete movement, or part skill, and the sequence to these movements. While there is some disagreement, most people recommend that the instruction proceed in the same sequence as the total motor skill. That is, you first teach the initial part of the movement, then the second part, or step, in the movement, and so forth. The student thus learns the part skills and the sequencing of these part skills at the same time.

Depending on the learner's mastery of the various part skills, you may provide additional practice on specific part skills that are not yet mastered. In that way, the learner would be practicing the precise part skills needed and not spend extra time on skills that had been mastered previously. Of course, after such practice of needed part skills, it is desirable to have the learner practice the total skill again so that he or she reinforces the learning of the sequence required in executing the motor skill.

Attitudes. The sequencing of instructional content in the domain of attitudes is not quite so straightforward. Due to the nature of attitudes it is not possible to find such tight knowledge structures that readily imply instructional sequencing. In general, it seems that an effective sequence for the teaching of attitudes is to first present a person the learners respect or with whom they can identify. Then this person is shown in a choice situation related to the desired attitude. They can choose to put on safety goggles (reflecting the desired attitude) or not. The person selects to make the desired choice and is shown receiving reinforcement from that choice. This progression of: (1) showing a high-status person, (2) in a choice situation, (3) making the desired choice, and (4) receiving positive reinforcement or satisfaction from that choice is the desired instructional sequence for the teaching of attitudes.

In summary, there is not one instructional sequence that is effective for all types of instructional outcomes. Learning is more complex than that. Different types of instructional content, i.e., different domains of learning, require different instructional sequences. In the remainder of this chapter we consider the sequence of different components of a lesson. These components, called instructional events, should be sequenced in a precise manner to facilitate learning by stimulating the cognitive processes operating in the memories of learners.

Sequencing Instructional Events

Before rushing into the development of an educational/training program it is important to consider what constitutes good instruction. What kind of things should a person developing the instruction build into the instructional materials to ensure that they do in fact *instruct*? What must the manager of a program look for in the design of instruction so that he or she can be reasonably sure that the instructional aspects of the program will be successful? This question is at the very heart of instructional design and affects the development of all instructional programs.

If we, or the materials we develop, are to provide instruction that *works*, then we must ensure that the text or other media we provide carry an instructional message. It is not sufficient just to provide *information* to the learner; we must provide *instruction* to that learner. "Telling" someone something is not the same as "teaching" someone something. We can all recall situations in which we were lectured to or were told something, yet we failed to learn what was intended. If teaching were the same as telling, we would all be brilliant teachers! Not all of what we read, see, or hear suffices to instruct us. If any communication that is intended to instruct, however, has certain features or characteristics, then that communication is much more likely to have instructional value. The purpose of this section is to examine these features, called *instructional events*, that turn information into instruction. Within the field of communications some researchers call this sub-area "instructional communications" to specify that it is a special form of communications. The topics treated in this chapter are often placed under the heading "instructional message design." While the terminology may vary, the intent remains constant—finding out what should be incorporated into a communication or message so that it instructs. These instructional events are based on much research on human learning, particularly within cognitive psychology. The instructional events are designed to facilitate the processing of information that occurs in our brains (Gagne, 1985). Since these instructional events are based on how we process information to learn, they are appropriate for all learning situations and outcomes. The following instructional events are based on the work of Gagne (1985).

Describe objectives. The first instructional event that should be incorporated into lessons is to include a description of the *intended outcomes* from the instruction. It is important that each learner knows exactly what is expected of him or her when he/she participates in a course. Perhaps the best way to do this is to use the objectives that were written when the lesson was being developed. The objectives establish the target we seek in developing the training program and individual lessons. These objectives can also be used by the learners to help them know what is expected as a result of going through the lesson. Thus, when a learner begins a particular lesson, he or she knows what will be learned and can focus attention accordingly. Successful learning does require that the learners direct or focus their attention on the relevant aspects of the instructional stimuli. By providing a statement of our objectives, we may be able to assist the learner in doing this (Gagne and Briggs, 1979).

Provide content overviews. The second instructional event that should be incorporated into lessons is to provide an *overview of the content* to be acquired. This is especially important when teaching information. The purpose of an overview is to provide a roadmap showing the "lay of the land" before a learner begins a journey through the material. The objectives describe what the learners will be able to do following successful instruction; this content overview describes how the instructional content itself is organized. This is the "big picture" with which instructional materials should begin (Reigeluth, 1983).

This technique of first presenting the essential features of a body of content in an overview, then gradually providing more detail has been termed the elaboration theory of instruction (Reigeluth, 1983). While the theory itself is complex, the basic idea is simple—present the main ideas first, then fill in the detail. Remember when you found yourself in a new city and quickly had to find your way around. Most people try to learn the major streets first, the main north-south and east-west arteries. Then later they fill in the side streets and, finally, the alleys and smaller by-ways. The major arteries are our content overviews. Present them first to give the learner the basic structure on which the more specific content can be anchored (Ausubel *et al.*, 1978). In fact, it may be desirable to allow learners the option of returning to the overview so that they can mentally keep track of where they are in the instruction and thus better integrate general and specific knowledge. That way, learners would build knowledge structures in their memories that were better organized and more resistant to forgetting.

Provide advance organizers. The third instructional event is similar to the second and has the same theoretical roots. This event is to provide an *advance organizer* for the student. An advance organizer is a brief introductory passage presented prior to the instructional material that has the purpose of assisting the learner to understand how to organize the instructional content and relate that to what he or she already knows (Ausubel *et al.*, 1978). Advance organizers are at a higher level of abstraction than the material that they precede. This type of instructional sequence is a whole-to-part sequence, since the most general ideas are presented initially. As such an advance organizer differs from an objective, since objectives refer to what the learner will be able to do following instruction, and the advance organizers provide an introduction to the material that is relatively abstract and general (Fleming and Levie, 1978). The advance organizer should help learners figure out how and where to relate the new information in their current store of knowledge. Also by facilitating the use of comparison of the new information with existing knowledge, the acquisition and retention of the new material should be enhanced. Learners are encouraged to compare the new, unfamiliar information with the information in their memories that is well understood. While this description of why advance organizers work is somewhat speculative, since we can't look into a learner's head to see these things at work, research on advance organizers has documented the positive effects on learning that they produce (Mayer, 1979). While there are interesting theoretical discussions regarding the effects that advance organizers have on the learners' cognition, the important thing to remember is that if we build advance organizers into lessons, our learners are more likely to be successful than if we don't.

Provide examples. The fourth instructional event is to *provide examples* demonstrating what is to be learned. Rather than a rambling description about what a right triangle is, provide the learner with examples of them. Unfortunately, there are often some surprised learners when, after a lecture or written passage on some new concept, they encounter an example of an incompletely learned concept. Such was the case in a computer fundamentals class when after hearing a description of light pens and viewing chalk drawings of light pens, the learners actually saw a light pen yet few could identify it! If we are to learn new concepts, then the use of examples is crucial.

Organizing and Sequencing Instructional Content 135

Describe prerequisite knowledge. The fifth instructional event to include in lessons is an indication of the *prerequisite knowledge* the learner is expected to have before beginning a particular lesson. Many times whether the learner possesses this prerequisite knowledge determines the extent to which the lesson is effective (Gagne, 1985; Gagne, Briggs and Wager, 1988). This is particularly important in the domain of intellectual skills in which new skills build upon existing skills. Before someone can be successful in learning one of the higher order skills, he or she must have mastered the lower level prerequisite skills.

In developing lessons we should remind the learner of these prerequisite skills. This does not have to be very elaborate, just enough to *stimulate the learner's recall* of that skill. For example, if the lesson was to teach the learner how to find the surface area of three-dimensional objects such as a cube, you could stimulate the recall of prerequisite knowledge by reminding the learners that they have learned how to calculate the areas of simple squares by multiplying the length by the width; then show them an example of this. Then in the lesson on surface area of cubes you could build upon knowledge of calculating the area of rectangles. Sounds simple! Actually, learning complex intellectual skills is indeed remarkably simple when the prerequisite skills have been mastered and their recall is stimulated.

Provide elaborations of the content. The sixth instructional event is to include *appropriately sequenced elaborations* of new content. The idea here is the same as presented earlier in the discussion of the use of content overviews and advance organizers. The sequence of the presentation should be from whole to part, general to specific (Reigeluth, 1983). This differs from programmed instruction, which was designed in a linear sequence of very small steps of question-response-feedback frames. This concept of instruction was based on the principles of behavioral psychology. According to this view of learning, the instruction must go slowly from specific to general, from part to whole. Many lessons continue to be developed accordingly. The research evidence that began to be accumulated in the 1970s seriously questions this view (McKeachie, 1974). A more appropriate sequence is the general-to-specific sequence recommended by elaboration theory (Reigeluth, 1983).

Provide prompts and guidance. The seventh instructional event is to *provide prompts and guidance* for the learning. When presenting new material to the learners and expecting responses from them, make liberal use of devices to prompt or cue their responses. This is particularly important for the early responses. Rather than let the learner spend much time guessing what response is required, provide a cue for the response. Don't expect the learners to have immediate total recall of new material, give them a little assistance. Present strong initial prompts then gradually withdraw them.

Provide exercises. The eighth instructional event is to *provide exercises* or problems so the learners can practice using the new content. Like the use of prompts, the provision of exercises is a holdover from programmed instruction that seems to be effective in promoting learning. Exercises or test-like situations allow the learners the opportunity to practice recall of what they have been learning, and the practice of such recall promotes retention. To be effective the exercises should be interspersed throughout the material and not bunched together at one point. As the lesson proceeds, the exercises might increase in difficulty to reflect the progression in content. Of course, the exercises must be directly related to the objectives of the lesson if they are to have the desired instructional value.

Provide feedback. The ninth instructional event is to *provide appropriate feedback* to the learners following their responses to exercises. The use of practice alone is not sufficient, the learners must receive feedback on their practice. Ideally the feedback would come right after their response, a situation very unlikely to occur in most classrooms but common in some forms of instruction such as CBT. Consider using informational feedback rather than a simple correct/incorrect. Explain *why* the learner's response was incorrect. If done correctly the provision of feedback can be one of the strong points of any lesson.

This concludes our description of the instructional events that give rise to learning when incorporated into lessons. When developing a course, keep these things in mind and put them into your lessons to ensure that the learning will be successful. When planning or evaluating courses it is appropriate to examine the lessons to determine whether these instructional events were included and how well they are used.

Summary

In this chapter we have described how to organize the content of education and training programs into courses and then lessons. We discussed how to analyze the objectives of a lesson by identifying the domain of the objectives and then using the appropriate learning task analysis method. We discussed how to sequence the instructional content of lessons. Finally, we described the nine instructional events that should be built into each lesson. More specific information on these topics can be found in instructional design books.

Chapter 10

Methods and Media Selection

All training and educational programs require a systematic determination of what combination of methods and media will best operationalize the training design. The decision has an impact on both effectiveness and cost. Once the design is complete, these decisions must be made before the actual development of courses and materials can begin. The design describes *what* is to be accomplished, while selected methods and media determine *how* to accomplish it. This choice is generally made on the curriculum level followed by a more specific determination for an individual course.

Research tells us that there is no one best method or medium. Thus, there are no inherent features about any one approach that make it better or worse than another. Decisions about what methods and media to use are driven by the objectives and instructional strategies established during the design phase. That is, for certain types of objectives, conditions and events, certain methods and media will be better than others. This criterion is the basis for an initial determination and generally narrows the selection to several alternatives. This first set of options must then be evaluated in light of existing resources and constraints. Yet, the rationale for decisions about how to deliver the training often includes such statements as "we have always done it that way" or "it is the cheapest way" or "we have got all this equipment sitting around and need to do something with it." In recent years, a major pitfall has been our love affair with high technology. Many people think that what is new and more technologically advanced is better than simpler or more traditional forms of instruction. Wilbur Schramm reviewed the effectiveness of instructional programs that were mediated through high technology and those instructional programs that relied on simpler forms of media on a world-wide basis in his book *Big Media, Little Media*. He wrote that television, films, and computers are not "super" media; they are simply "spectacular media" (1977:275).

This chapter will first describe the characteristics of the various choices of methods and media, followed by a discussion of criteria to use in selecting the best option(s).

Feasibility Analysis

Decisions about methods and media are generally made on the curriculum level first. Once a series of training topics and the target audiences are identified, a range of feasible options for delivering the instruction are analyzed. At this point, the training designer generally knows the following:

1. What training topics are needed.
2. Who is the target audience(s).
3. What resources are available.
4. What constraints must be considered.

Knowledge of the training topics informs the training designer about the kind of skills that are associated with the material. Is it, for the most part, manual procedures that may require, for example, a number of demonstrations; or, might it be more intellectual and analytical in nature and thus require opportunities for discussion and problem solving? The designer also has some idea of the target audience. He knows, for example, if a given group has a limited or advanced understanding of the material to be covered. Such information will determine if, let us say, lectures are appropriate. Such can be the case with introductory information. On the other hand, if case exercises are incorporated into the training material, this may require more advanced knowledge of the topic. He should also know about their past training experiences. Has this group had experience with more innovative methods, such as role playing and group discussions, or are they only accustomed to traditional lectures? If a group traditionally associates education with lectures and library research, they may have difficulty grasping the educational value of group discussions early on. Additionally, the designer should know what kind of personnel, financial, and logistical support is available. Also included is knowledge of deadlines. Thus in deciding what methods and media are possible, the designer needs to know if the time, money, and support are available to design, for example, self-study instruction for videodiscs, or if a simpler, cheaper, and quicker method and media combination is more reasonable.

Thus a feasibility analysis simply states what methods and media constitute a reasonable set of alternatives to deliver the training. Can the instruction be conducted via lectures and small group discussions, or can it be designed in a self-study format?

Feasibility analyses are a fairly general type of evaluation. They are based on the overall types of objectives that are to be achieved, the conditions of learning they represent, and the existing resources and constraints associated with the training environment.

Description of Methods and Media

Methods

There are eight basic categories of instructional methods that can be used to operationalize a training or educational design. These are the traditional lecture, large or small group discussion, tutorial, independent study, demonstration, role play, simulation, and on-the-job training.

Many instructional strategies use a combination of methods and media to deliver the training. For example, the self-study method can be enhanced with mentoring assistance or small group seminars or via informal discussions. This is because people can generally learn more when they have the opportunity to interact and receive feedback from others. Yet a combination approach that includes or even stresses the self-study method may be required in this instance because interests and time available to learn are highly varied among learners.

Lecture. While lectures can be presented before any size group, they are typically associated with presentations before large audiences. Generally, a speaker, usually a content expert in a particular field, talks to the audience for a period of time on a particular topic. This method is for the most part a form of one-way communication and is one of the most widely used and often abused forms of delivering instruction. The audience listens but does not interact. In turn, the lack of interaction between the speaker and the audience promotes passive learning, which can lead to lower retention with regard to both short- and long-term memory. It is also heavily instructor-dependent, which often results in variance in instructional quality. Often lecturers do not customize their lectures for individual audiences and may therefore be less responsive to the individual needs of different groups.

The lecture mode can be enhanced if the speaker asks questions to make the audience participate more and think about the ideas or concepts. Lectures are good for presenting general information, and are economical in that little or no development time is required and time is spent on discussion.

Discussion. In the discussion method, the trainer and/or participants exchange ideas about a given topic. This is a more active form of learning and is better for teaching analytical skills than the lecture method. Here virtually everyone participates. Discussions can employ structured group techniques such as the NGT to facilitate brainstorming, or they may work together on a less formal basis. Often groups are assigned a problem to solve or an exercise to complete; they operate more as a task force that may deal with issues that are real or imaginary.

While discussions can occur in a large group format, they generally work best with smaller numbers of people. The best size may be from five to seven (Stater, 1958; Weaver, 1981). This number permits participants an opportunity to be actively heard on a regular basis as well as time to reflect on the topic at hand. Groups with fewer than five members force individuals to constantly participate to maintain the discussion, while group sizes of more than seven make it difficult for all to be heard on an equal basis.

Group discussions can encounter problems when participants are not accustomed to this kind of learning or may be afraid to speak out because of certain relationships with other group members. Thus it is often best not to mix staff and supervisors, for example, unless the objective specifically calls for them to mix to promote a greater understanding of each other's perspective on given issues and problems. When appropriate, it is a good idea to encourage groups to pick a discussion leader. Often this process suffers if there is not enough direction, as participants may easily diverge from the topic at hand and require guidance to maintain desired focus.

Tutorial. The tutorial method is a one-on-one relationship between the trainer and the trainee. Instruction often consists of informal discussions and question and answer sessions. It is often used in combination with other methods, such as independent study. The individual attention inherent in this approach is especially useful in providing remedial or catch-up training. It permits an opportunity for clarification via examples that are unique to a given student as well as an opportunity for close monitoring and immediate feedback from the instructor.

Independent study. Independent or self study is individualized instruction. Participants may use a set of guidelines to seek out their own material or they may participate in a form of programmed instruction. Both self-study techniques are self-paced, flexible, and generally available on demand. Programmed instruction may come in the form of printed self-instructional materials or teaching machines. Computers are commonly used to mediate this type of instruction.

Programmed instruction permits participants to repeat instruction until they master it. However, this type of repetition combined with the lack of interaction may cause some participants to consider this method boring or juvenile. An additional disadvantage is the high cost associated with the initial development of programmed instruction. This cost may, however, be amortized over time.

Demonstration. This technique is excellent for instruction that requires the learning of step by step procedures, such as changing a tire, operating a piece of machinery, or completing a set of forms. The demonstration method can be used by the trainer to illustrate the correct way to execute a performance, or it can afford the trainee an opportunity to practice and receive feedback on his or her own abilities. Demonstrations are often combined with other methods when intellectual skills are required to determine when to best employ a given set of procedures.

Demonstrations are best conducted with small numbers of people. This format permits the best visual and hands-on opportunities. Sometimes demonstrations can be expensive to operationalize. They can be costly with regard to both props and mistakes and this is especially true where safety conditions are a concern, such as learning to pilot an airplane or learning to drive a car.

Simulation. Simulations can be used to teach both procedural and problem-solving skills. For the teaching of procedural skills, simulations can provide a cheaper and safer environment than demonstrations. With this method participants can practice behaviors and receive feedback via models or mock-up situations. Computers, machinery, and other specially designed equipment can be designed to teach people to fly aircraft, drive cars, and operate machinery.

Problem-solving skills can be taught via case studies, in-baskets or simulation games. The case study method invites participants to seek solutions to oral or written accounts of a conflict situation. Interaction is fostered through the shared experience of analyzing problems together. Case studies can be both expensive and time-consuming to prepare or purchase. Commercially prepared cases need to be carefully evaluated, as they are often criticized for not containing enough information. Simulation games are like case studies. They, however, have the additional element of competition among other small groups who deal with the same problem or issues. Finally, in-baskets give participants an opportunity to practice managing and setting priorities for problems and other organizational materials such as phone messages and memos typically found in a carefully arranged desk-top in-basket. This technique is used to teach skills needed to delegate, schedule, and plan how to carry out tasks.

Role play. Role plays permit participants to learn by pretending. Roles that approximate real life situations may be acted out individually or as a group. Role playing can be designed to give people the opportunity to make new behaviors habitual through practice. Participants may also be asked to reverse roles which can lead to greater empathy for the perspectives of others. A classic example is when a husband acts out the role of his wife and vice-versa. This method is good for teaching interpersonal and practical skills in areas such as counseling, interviewing, customer relations, effective selling, or conflict management.

While this technique is very effective in helping people to recognize that there is seldom one best solution to any problem or conflict, it can be a time-consuming method. Participants may additionally feel uncomfortable if they are unfamiliar with the role or unaccustomed to expressing their emotions in front of other people. Role-playing may also be perceived as artificial, as is sometimes the case with simulations.

On-the-job training. This method permits trainees to learn how to do assigned tasks without leaving their jobs. Typically, individuals learn required skills as needed from experienced staff and supervisors. There is little wasted time associated with this technique, as instruction is individual, on-demand, and trainees learn only what they need to complete the job at hand. It is an inexpensive method to employ, as there is practically no advance preparation or administration cost other than ensuring that enough slack time exists in deadlines to permit this kind of learning.

This technique does not, however, permit trainees the luxury of practicing their skills in a safe environment. Also, the quality of instruction may vary, as it is highly dependent on the time, availablility, and communication skills of those personnel assigned to mentor less experienced staff. For this reason, it is best used as an opportunity to practice and refine skills learned elsewhere, and is seldom recommended for the initial teaching of new skills and knowledge.

Key Methods—Advantages and Disadvantages

LECTURE
Advantages
Time saving; can deliver content to a large number of people in a short amount of time.
Good for presenting general information.
Can be used with any size group.
Disadvantages.
Instructor-dependent.
Quality can vary.

One-way communication leads to passive learning.
Can lead to boredom, limited attention span and lower retention rates.
Lack of feedback for the instructor.
Presentations may be "canned" and unresponsive to specific audience needs.

DISCUSSION

Advantages

Involvement of everyone.
Reduced peer pressure.
Exchange of ideas and shared experiences.

Disadvantages

Rests on voluntary participation.
Limited effect in large group formats.
Some participants may be unfamiliar or uncomfortable with the process.
Groups can easily digress when there is no leadership.
Time-consuming.

TUTORIAL

Advantages

Individual attention and customized instruction.
Generally self-paced.
Good for providing remedial training.
Close trainee monitoring and immediate feedback.

Disadvantages

No interaction with others who have similar learning needs.
Labor-intensive.

SELF-STUDY

Advantages

Self-paced.
Immediate feedback with programmed instructional formats.
On-demand training.

Disadvantages

Lack of group involvement.
The repetitive nature of programmed instruction may be fatiguing or perceived as juvenile.
May be time-consuming to design and develop.

DEMONSTRATION
Advantages
May be used by the trainer or the trainee.
Permits hands-on experience.
Opportunity to try out new performances in real life settings.
Immediate feedback.
Disadvantages
Mistakes can be costly and dangerous.
Time-consuming.
Labor-intensive.
Ineffective with large groups.

SIMULATION
Advantages
Opportunity to practice in a safe environment.
Can be used to teach procedures as well as problem-solving.
Theory is made tangible through actual performance.
Perceived as fun and challenging.
Can be used to enhance team work skills.
Provides opportunity to see problems from other perspectives.
Disadvantages
Needs to be very carefully planned to achieve desired objectives.
Time-consuming to administer and design.
Can be expensive.
May be perceived as artificial.

ROLE PLAY
Advantages
Opportunity to practice new behaviors.
Opportunity to reverse roles and see problems from other perspectives.
Good for the teaching of interpersonal skills.
Disadvantages
Time-consuming.
Ineffective in large groups.
Participants must be accustomed to expressing emotions in public.

ON-THE-JOB TRAINING
Advantages
Responsive to on-demand training needs.

Participants learn only what is needed.
Participants can remain on the job.
Low administrative costs and requires little preparation.
Disadvantages
Less effective for initial teaching of knowledge and skills.
Rests on ability and willingness of experienced personnel to train others.
Instructional quality may vary.
This type of training may suffer if the work effort is behind schedule.

Media

Once the method(s) have been chosen, media are evaluated to determine which one(s) will best support the selection. For any given method there is generally more than one option or combination of media that will help deliver the training. Media can be visual or audio or both visual and audio. In turn, media can also vary by being either static, transient, or computerized. There are four basic categories of media. These are print, visual aids, transient media, and computerized media.

Print. Print includes the use of books, manuals, programmed texts, and microfiche and microform. This medium requires materials to be written for the appropriate reading level of the target audience. In turn, all participants should possess similar levels of literacy. This medium is portable, relatively inexpensive to use and can be produced easily in mass quantities.

Visual aids. Like print, visual aids constitute a static medium. They include the use of charts, diagrams, graphs, illustrations, drawings, photographs, exhibits as in models, and projected still images such as opaque projections, overhead transparencies, and slides. Visuals are effective because people *expect* to see them. Since the advent of TV many people have become more visual-minded and may tend to expect to see visuals as an illustration of ideas and concepts. Visuals also enhance retention. Visual aids are generally used to illustrate verbal information and range from concrete drawings to abstract symbols. In developing countries visuals are often used to transcend literacy barriers. As in all cases, audience characteristics need to be taken into account.

For example illiterate populations often view cartoons as condescending. In addition, groups who are not accustomed to seeing the enlargement or reduction of items may not recognize familiar objects and misunderstand the purpose of the instruction.

Transient media. Media that are transient are both visual as well as audio. Transient media that are strictly audio include radios and tape recorders. Visualized media in this category are filmstrips and silent motion pictures. Transient media that are both audio and visual include slides, motion pictures, and filmstrips with sound, TV, and the motion picture as a repetitive loop.

Transient media are often used to broadcast instruction as with the TV based Open University in England and the University of the Carribbean's correspondence study program through radio. These media can be used to demonstrate performances before large audiences and emphasize certain points through close-ups and slow motion. In turn, they are an essential choice when there is a need to visually demonstrate time, motion, and social interaction. These media are of course more expensive to produce and require the use of electricity as well as sophisticated hardware and a darkened room, which may not be available in all training situations.

Computerized media. Computers are synonymous with high technology and have achieved a high degree of popularity as the number of personal computers and computer terminals in homes and offices has increased dramatically. Computerized instruction includes CAI, that is, the use of the computer to teach, and CMI, the use of the computer to manage instruction by keeping records, by pre-testing trainees out of certain portions of the training sequence, and by prescribing remedial instruction as needed. This category also includes the use of interactive video and videodiscs to teach interpersonal, counseling, and management development skills.

They can be used independently, when needed, and permit greater control over the process and quality of the instruction. That is, learners can control both the amount of time needed to complete a lesson and the need for assistance, e.g., more examples or another look at the glossary. Perhaps one of the computer's greatest instructional advantages is its branching capability, which permits learners to try out more than one solution to a given problem. In turn, students are able to see how the consequences of their actions can vary. Most trainees perceive computerized instruction as motivational. Trainees also find that they require about one third less time to learn a given lesson. However, it is important to note that CAI works best when it is combined with other instructional methods and media, as most individuals need additional interaction and feedback from instructors and fellow students.

The use of computers also facilitates the decentralization of instruction at a time when a lot of organizations are concerned about the high cost of travel and the lost opportunity costs associated with centralized programs. Its benefits are best realized when there are a large number of people to train in different geographical areas and when the instructional content is stable.

While the administration of this kind of media is less labor-intensive than traditional instruction, design and development costs for computer-assisted instruction can be at least three to six times greater. The design of interactive videodiscs can require as much as 30 to 40 times more effort. Computers are often used as expensive page turners if the necessary effort is not given to course design. The worst case scenario and the least cost-effective is to use computers for drill and practice. This kind of instruction is sometimes called the "flash card" approach, as it simply presents a question followed by feedback for right and wrong answers. Costs, however, are a one-time commitment and can be amortized over time as more and more trainees are trained. The inverse is true of traditional classroom instruction.

Key Media Advantages and Disadvantages

PRINT
Advantages
Can be used independently at a trainee's own pace.
Inexpensive to produce.
Easily portable.

Can be used as reference materials after the training.
Disadvantages
Requires all trainees to possess similar levels of literacy.
Must be translated into other languages when used abroad.
May be fatiguing if used alone or for a long period of time.

VISUAL AIDS
Advantages
Effective for illustrating verbal information.
Portable.
Can transcend linguistic and literacy barriers.
Modifications are relatively easy and inexpensive to make.
Disadvantages
Ineffective for large groups unless projected on a screen.
Visuals that do not approximate reality may be perceived as confusing.

TRANSIENT MEDIA
Advantages
Can use with any size audience.
Can provide examples that more closely approximate reality.
Can combine both sound and visuals.
Can demonstrate a performance before large audiences.
Can be used to show close-ups and slow motion.
Good for teaching interpersonal skills.
Can be used to broadcast instruction and reach participants in remote regions.
Permits control of instructional quality.
Disadvantages
Not interactive—promotes passive learning.
Not easily portable.
Requires a darkened room.
Can be expensive to produce and deliver.
Modifications are expensive and time-consuming to make.

COMPUTERIZED MEDIA
Advantages
Can be used independently and self-paced.
Permit repetition until content is mastered.
Immediate feedback.
Costs can be amortized over time.

Can be designed to be interactive.
Permits learners to "think through" more than one solution to a problem.
Can be used to manage instruction.
Perceived as fun.
Permits decentralized training.

Disadvantages

Design and development costs can be very high.
Modifications are expensive and time-consuming to make.
Content needs to be static.
Lack of interaction with other participants or trainer.
Limited ability to teach attitudes.
May be used as an expensive page turner if not well-designed.

Research on Methods and Media Effectiveness

Considerable research has been conducted over the past two decades on the area of methods and media effectiveness. Several general conclusions can be drawn from the research evidence.

There is no overall superior method or medium. The failure to find an overall superior method or medium has been consistently supported by research studies. The evidence does not support the notion that any method is significantly better than another (Schramm, 1977; Jamison, Suppes, and Wells, 1974; Clark, 1983; Gagne, 1985). Thus, the demonstration method is not better (or worse) than the lecture method. The research on media supports a similar conclusion. A videotape is not better (or worse) than a book with illustrations. So any attempt to select an instructional method or medium on some presumed overall superiority is without support.

The selection of methods or media should be based on a consideration of instructional strategy. Instructional strategies (objectives, conditions, and instructional events) constitute the heart of the training design. A common research finding supports the idea that certain method and media options are more effective in teaching some objectives than others. Additionally, much research has demonstrated the superiority of certain instructional conditions in the teaching of certain types of learning outcomes.

Methods and Media Selection 153

For example, it appears to be more effective to demonstrate the procedures for starting a car than it is to present a lecture on the topic. This research evidence can be best utilized to select appropriate methods and media by starting with the objective followed by an identification of the instructional conditions and events that are to be operationalized (Gagne and Briggs, 1979; Gagne, Briggs and Wager, 1988).

Matching the medium to the learner does not seem to have a direct effect on learning. Research has shown that the media preferences expressed by people do not affect their learning. Some individuals may prefer to read about a topic while others prefer to listen to a lecture on the same subject. Research implies that individuals are as likely to learn from a book as a lecture.

One conclusion is that most people (80%) are not media sensitive; they learn equally well from any medium. Another conclusion is that although some people express a preference for a specific medium, they will learn just as much fron a nonpreferred medium.

Visualizing information can increase retention. We have all heard that a picture is worth a thousand words. Research studies offer similar conclusions and indicate that the learning process can be enhanced by visualizing verbal information. It is estimated that 83% of what an individual learns is learned through sight (in Donaldson, 1985). It thus appears that the use of visual aids as either static (posters or slides) or transient media (films and videos) can make a difference in how well people learn.

Cultural differences can affect how well visuals are understood. Research indicates that individuals have certain expectations about the way certain images should look. When these differences are not taken into consideration, they can lead to message distortion, a longer learning period or simply a lack of learning (Rice, 1980; Rogers, 1973). For example, a visual containing a woman with short hair and dressed in slacks may be rejected or interpreted to be a man by audiences where short hair and slacks are considered culturally inappropriate. The same is true with transient media where individuals may have certain expectations about the pace, emphasis, close-ups, and the progressive flow of visualized events. (Angrosino, 1976; Salomon, 1970).

It may also be that the degree to which cultural differences affect or interfere with learning is associated with schooling (Hansen, 1985). Research evidence indicates that literate people are more likely to try to understand imagery that is culturally foreign while non-literates have a greater tendency to reinterpret or ignore images that are not a part of their immediate world.

The use of computers can reduce learning time. Studies indicate that the use of computers can reduce the time it takes to achieve a given set of objectives by as much as 30% (Kulik, Kulik, and Cohen, 1980). The research, however, is inconclusive about the ability of computers to increase actual learning. It thus appears that for certain types of learning, computers can help individuals learn quicker, but not necessarily more. Research studies also suggest that computers can be motivating. It appears that most individuals like to use them, which may also contribute to the computer's ability to increase learning rates.

Selection Criteria

The decisions regarding methods and media are based upon many factors, the most important of which are the objectives to be taught, the desired instructional conditions and the specific instructional events. This first set of criteria is followed by decisions based on available resources and existing constraints. When resources are not available to support the most desired method and media approach, the substitution of another option becomes necessary. Research has shown that often this "trade-off" does not significantly lessen the effectiveness of the instruction as long as the substituted approach will serve the specified instructional strategy.

To systematically select the best set of options to use, the following issues need to be considered:
  1. Objectives.
  2. Conditions and events.
  3. Time and money.
  4. Familiarity.

Methods and Media Selection

5. Size of group and frequency of instruction.
6. Format.
7. Assessibility, durability and convenience.
8. Ease and speed of production.

Decisions that are driven by the objectives, conditions and events specified in the instructional strategy are supported by the research that indicates that nothing inherent exists in any one method or medium that makes it better or worse than another.

When resources are not available to support the desired method and media approach, the substitution of author option becomes necessary. For example, the first choice may include a color video tape, but the client does not have the money or time to support this kind of design effort. You may instead wind up deciding to use a slide/tape presentation, because this is all you have the resources to choose. The same is true of environmental constraints such as the lack of electricity. Photographs, for example, can be used to replace films in remote regions. Research has shown that often this "trade-off" does not significantly lessen the effectiveness of the instruction as long as the substituted approach will serve the specified instructional strategy.

It is also important to consider the overall cost-effectiveness of options. Some material initially may be more expensive to produce, while costs may be equalized in the long run. This is true with computer-based instruction when a large number of people are to be trained on an ongoing basis. If, however, the course is to be taught only once to a small number of people, it may be more cost-effective to design a more traditional lecture and discussion presentation than an elaborate set of self-study materials.

Limitations on instructional format can also affect what methods and media will be more effective. When it is difficult to disengage individuals from their jobs and bring them to a centralized training session, it may be more efficient and effective to decentralize the training by providing self-study materials. In turn, lectures may be more effective when large groups are assembled, just as role playing or simulations games are likely to work better when individuals can be trained in small groups.

It is also important to consider the prior familiarity that intended students and instructors have with the options under consideration. If the choice has been narrowed down to two options based on what is necessary to convey the instructional strategy, it will be more effective to select those methods and media that are more familiar. A second factor is the training designer's familiarity with the selected methods and media. If the designer has never designed programmed instruction it will take more time and effort. Decisions about the designer's familiarity should, however, be secondary to the familiarity concerns of trainees and instructors.

Finally, it may be important to consider the ease and speed of production for various method and media options. This is especially true when deadlines are short. Likewise, in-house production capabilities are important. Not all organizations have the capability of producing all available media. It is generally more cost-effective to first review the availability of existing materials that may have been used in earlier training programs or the possible purchase of vendor materials before the decision is made to produce new, original materials.

Summary

This chapter has dealt with the selection of methods and media. This choice operationalizes the training design and must be determined before the development of course materials can begin. Research studies indicate there is no one best method or medium. Studies do, indicate, however, that certain methods and media will deliver the objectives and instructional conditions specified in the design better than others. Decisions based on this criterion generally yield several options for the selection of methods and media. Choices are narrowed down and a final determination is made on the basis of time, money, and other environmental constraints, size of the intended audience, planned frequency of instruction, prior familiarity with selected methods and media, the instructional format, and ease and speed of production. The decision-making process generally occurs first on a macro-level to make decisions that affect the overall curriculum. The selection is further refined during the design and development of individual courses.

Chapter 11

Project Management and Consulting

Instructional systems development efforts require considerable time and resources. Even moderately sized ISD projects can be complex because they involve several people doing different, but interdependent, tasks that must be completed in a timely fashion. It is quite possible for a project to go astray even though it was staffed with qualified people who knew how to carry out the steps in the ISD process. Effective instructional systems development projects require careful management throughout the whole process. Without such careful oversight and coordination, projects fail. Some of the needs for management are not at all unique to ISD projects; any time there is a group of people working on a variety of tasks over time, there is likely a need for management. When the tasks are interdependent, as in instructional systems development projects, then the need for management is even greater.

It takes time, money, and people to do ISD well. Most often the individual who holds the purse strings within an organization is a functional person who will grossly underestimate the needs of an ISD project. This person is used to the "old style" of training in which a content specialist is given the task of teaching others. All the upfront planning and backend evaluation may seem a bit "trendy" for his/her pocketbook. It is essential to the success of the ISD project that this person "buy" into the process to ensure that the required resources are available. He or she must come to see that spending money now will save money over the long run. It is also important that management sign-off on each major decision, or at least on each phase, of the project. This will promote acceptance and also diminish liability in case of any problems.

We suggest that designers and not functional/content persons should direct ISD projects. In some organizations this arrangement may be a bit delicate. For example, a GS13 ranked person may be asked to take direction from a GS1 on a specific government project. It is essential to the success of a project that the designer have the required authority and leverage to make ISD decisions; that is, decisions driven by instructional principles. Of course, content persons and ISD specialists can discuss how to go about some task, like whether to use lecture or computer based training to teach a certain objective. However, when it is time to make the decision and move forward, someone must be empowered to say "This is the way it is going to be." That person should be a skilled designer, not a content person. Rarely is there time for content persons to learn enough about ISD to function in this role.

Instructional systems development projects by their very nature involve working with different types of people. A typical project involves several different kinds of personnel:

(1) people who are knowledgeable about the content;
(2) people who are skilled in instructional systems development;
(3) people with skills in producing the instructional materials whether computer-based instruction, instructional television, or printed materials; and
(4) a variety of specialists, such as computer programmers, graphic artists, or cameramen, as needed for a given project.

ISD projects are not the effort of an individual; these projects require different types of talent and skills, not often found in one person. Because of the diversity of people working on instructional systems development projects, and their professional allegiances, the manager of an instructional systems development project will need to be sensitive to the unique contributions of each individual and must spend time keeping the communications running smoothly. Effective ISD projects are a case in which the whole is truly greater than the sum of the parts.

How do ISD projects go astray? None are ever planned to fail; no projects are expected to result in a disaster, yet some do.

Projects fail when they are not staffed with people who have the knowledge and skills necessary to carry out the steps in an ISD model. A competent staff is necessary. This staff must be following some type of ISD model that includes the essential steps in the instructional systems development process. Of course, there is room for more than one "true" ISD model. A survey of the literature reveals many ISD models with remarkable similarities (Andrews and Goodson, 1980). There is a core of basic processes, or steps, that must be followed in order to reap the benefits that should accompany ISD. Many ISD models include this "basic set" of ISD processes. When there is an adequate staff following an acceptable model in completing an ISD project the remaining area associated with failure is inadequate management. If resources are not made available or are misused, if schedules are not kept, if the quality of the output from each step in an ISD model is not monitored, if effective communications are not the norm, or if a spirit of mutual respect and cooperation among the staff is missing, instructional systems development projects can go astray and be very unproductive. It is a management function to see that these situations do not occur and spoil what otherwise could have been a successful project.

This chapter describes the role of management in helping instructional systems development projects be successful. There are two major aspects to this task of managing ISD projects: (1) project management and control and (2) consulting/negotiating tasks. First we will look at project management and control and then we will examine strategies for successful consulting and negotiation.

There are three essential parts to effective project management and control. The first part is the ability to estimate, or plan, what must be done. This requires knowledge of the specific steps that make up an ISD process and the amount of time and resources that are required to complete each step in the ISD process. The second part to effective project management and control is the ability to schedule the work necessary for the completion of an ISD project. The third part to project management and control is the monitoring of the work during the operation of the project and making adjustments as necessary so that the project is completed on time and within budget.

Estimation. One of the very earliest tasks of a person managing an ISD project is the estimation of the time and resources required. This is true regardless of whether the project is funded externally or internally. Funds must be made available to cover the necessary costs of the project. Personnel must be hired or assigned to the project. The manager of the project must obtain authorization to initiate the project. All of these essential actions involve top management approval. Before granting that approval, top management will usually require a detailed plan for the project that includes such estimates as the amount of time required, the amount of dollars required, and the personnel needed to complete the project. Thus the place most managers begin when undertaking an ISD project is with careful estimates of what will be required to complete a project. Managers must be skilled at completing these estimates because if they underestimate the project, sufficient resources may not be available to complete it. If the estimates are unrealistically high, then authorization to undertake the project may be placed in jeopardy. In either case, poor estimates damage instructional systems development projects.

The amount of time required for certain steps in an ISD project is often underestimated. One area in which time is usually underestimated is for materials production. Managers of ISD projects rarely allow sufficient time for materials production. Another area often requiring more time than estimated is the review of certain products during the project. Most large organizations require a series of reviews to serve as a check on instructional materials. In addition to a review by content specialists, some organizations require a review by management, personnel, legal staff, or the publications department. All of these add to the time required for project completion.

So how does one go about becoming a better estimator of what will be required for completing a given ISD project? Experience with ISD projects is the best teacher of this. Clearly one must be familiar—quite familiar—with the steps in the ISD model being used. It is difficult to estimate how long something will take if you don't know how to do it. We don't have the benefit of a "flat rate" book such as automobile mechanics have that shows the amount of time common tasks take, such as how long it

Project Management and Consulting

takes to replace a defective fuel pump on a 1988 Chevrolet. In the absence of such standardized guidance, ISD managers must draw upon their own experiences. So experience becomes vital. One can talk with more experienced ISD managers in his or her company to find out what reasonable estimates are, given the company's history with ISD projects. Take the opportunity to tap the company's institutional memory regarding similar projects. While "flat rate" manuals don't exist, there are some ballpark estimates that float around among ISD managers. For example, within one organization that has developed much computer-based training (CBT) for their employees, it is generally acknowledged that something on the order of 400 person hours are required to develop one hour of good CBT. If the estimate of a manager for a new CBT development project varies greatly from this common estimate, that estimate will be questioned. Thus, it is vital that persons managing ISD projects explore the often unwritten norms within their organization. Estimation is a skill that most managers improve at over time.

When estimating a project, of necessity you will have to make certain assumptions about such things as how long certain activities will take or how much certain items will cost. Document these assumptions; don't just give a final figure but rather list each item and its cost in the contract or planning document. Plan for the time your client will spend during the ISD project. Build in time for review and approval where this is likely to be required. When consultation with outside content experts is going to be needed, allow adequate time for it. Since an ISD project is not done until an evaluation and revision of the first draft are completed, make sure you allow adequate time to collect and analyze evaluation data and to make resulting changes. When estimating the output of your staff, remember that we work at different rates. Some persons move faster than others and some people think faster than others. More experienced workers should be able to complete tasks sooner than novice workers. Use a realistic figure when estimating worker output, not what a person can do under ideal circumstances. You are estimating what *will* likely happen and not what *could* happen. When the estimate is done, add some time and dollars for reserve because the unexpected will happen! Many project estimators use 20% as a safe reserve.

Once estimates are completed and the project initiated, the allocation of time must be monitored. Rather than moving forward on a day by day basis, effective managers complete a schedule for all the tasks required for a project, how long each task will require, and when it must be done. Often such a detailed schedule is done as part of the project's estimate before the project is undertaken. While there are several ways to go about completing a project's schedule, two standard ways have emerged. These are a Gantt chart and a PERT network. Both are ways for scheduling the work to be done on a project. They differ in the specific techniques used.

A Gantt chart shows the relationship between a list of tasks to be performed, the amount of time required by each task, and when each task is to be executed. Figure 11.1 shows such a Gantt chart.

As you see, the left column is the listing of specific tasks that must be performed in completing a specific ISD project. Reading across to the right you see a beginning and ending time for each task, indicating how long the work will take. By reading upwards in the chart, you see on which days this task will be done.

Although some of the intent is the same, a PERT network or diagram is different. In completing a PERT diagram the project manager must identify all of the important events or milestones in a project. An event refers to the completion of some activity. For example, event number 4 may be completion of the terminal objectives for an ISD project, event 6 may be the selection of instructional media, and event 12 may be analysis of formative evaluation data. The event refers to the completion of some necessary aspect of the project. These events are displayed on a diagram that shows the interrelations among events by connecting lines. Figure 11.2 is an example of a PERT diagram.

PERT diagrams are read from left to right. The numbers along the lines that connect the events represent the amount of time necessary to complete an event. Thus, the number 6 between event 3 and event 4 indicates that once the 3rd event is completed, 6 hours are necessary to complete event 4.

Project Management and Consulting 163

Figure 11.1 Gantt Chart.

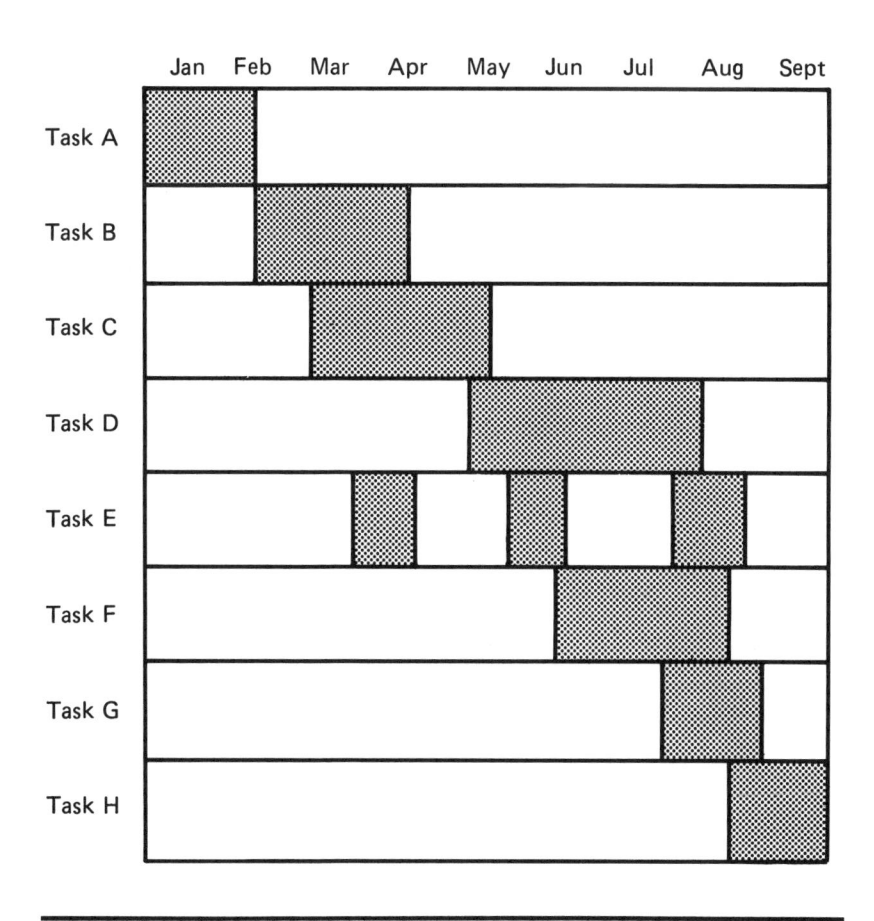

Figure 11.2 PERT Network.

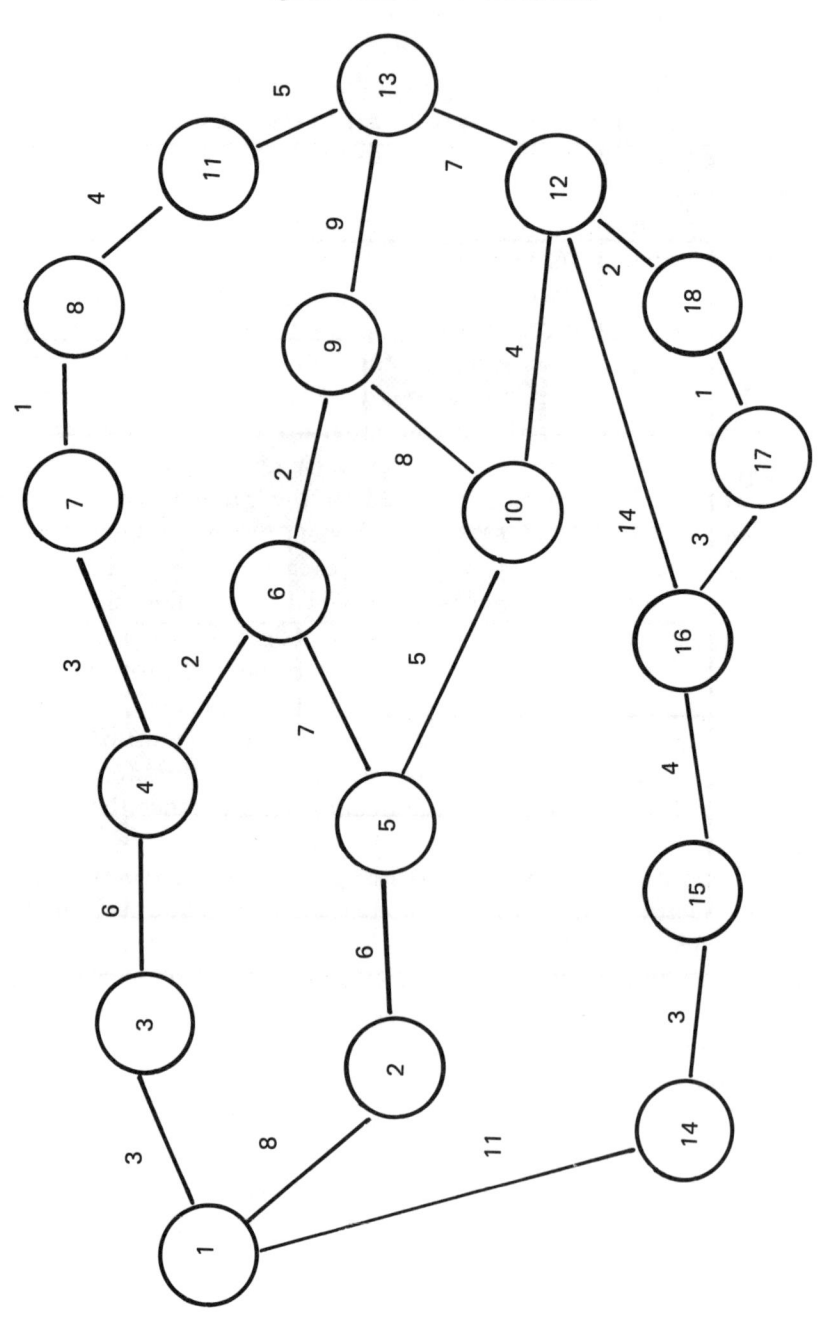

Project Management and Consulting

In developing a PERT diagram the manager is forced to think through an ISD project, identifying the events required, what events they depend on, and how long they take to complete. The analysis of the resulting PERT diagram will indicate how much total time the project will require. Once a project is underway, the manager can determine what will happen if some event is not completed as planned. Thus, PERT is also useful as a project is being worked on.

The final aspect of effective management of instructional systems development projects is monitoring of the quality of the work. A good ISD project manager must know how to examine the outcomes of each step within an ISD model. He or she must be able to determine whether the job analysis is acceptable, whether the objectives are adequate, if the instructional strategy is sound, and whether the instructional materials are appropriate. In short, an effective ISD manager must know how to evaluate the work of each person involved in the project. Perhaps the manager is not able to perform each step, but he or she must be able to determine the adequacy of each step. While a baseball manager might not be able to pitch and hit as well as the team's players, the manager should be able to judge the performance of his players. Likewise with the manager of an ISD project. He or she doesn't have to be able to conduct a task analysis, develop a CBT lesson, or complete an evaluation but he/she must be able to determine whether the task analysis was properly done, whether the CBT lesson is of acceptable quality, and whether the evaluation was completed successfully.

A quality control program is essential in ISD projects due to the cumulative nature of ISD models. Any mistakes made in the early steps of an ISD project, such as a faulty needs assessment, will be compounded as the project progresses. The development of effective and efficient training programs requires that each step in the process is examined as a project progresses. This is a key management responsibility.

A manager must have standards against which he or she will judge a project. The results of each step should be compared to the standard and found to be acceptable before the manager lets the project go forward.

In summary, instructional systems development projects are complex undertakings that require careful management if they are to be successful. Managers must estimate what will be required to complete an ISD project, schedule the work to be done, and monitor the progress along the way.

ISD as a team effort. As previously mentioned, no one does instructional systems development alone. There are too many tasks to be done, and these tasks require many different skills so that any one person would not likely have all the required competence even if time weren't an issue. Thus, ISD projects are staffed by many people with different professional backgrounds and different ideas about education and training. Establishing and maintaining smooth working relationships among these people on ISD projects is a task of the project manager. There are conflicting values and demands from the staff. The job analyst wants more time to complete a task analysis; the content specialist is ready to skip over the front-end analysis altogether since he "knows" what is important; the graphic artist is ready to create museum quality illustrations; the computer programmer wants to get started with the detailed coding and doesn't want any more changes during formative evaluation; the instructional designer is trying to develop an instructional strategy; and the technical writers are beginning to write. The manager must coordinate their work and get the best product he or she can from each person while not allowing any one person or group to dominate. Welcome to ISD management! Obviously a manager must have some very good "people skills" and must be a good judge to resolve disputes.

Conflict can arise between designers and media producers. This is particularly true when producers have not been trained in ISD procedures or do not have ISD experience. Designers select certain media because of the fit with the objectives. Producers link their selections to their knowledge and skill with the medium; often they have a favorite medium and are quite passionate about it. To designers, media production is part of a larger whole; to media producers, media production is often the end in itself.

A manager of ISD projects must be able to deal with problems that typically arise when working with content specialists to produce instructional materials. While an ISD project should afford people of varying backgrounds the opportunity to collaborate in creative work leading to high-quality products, it often provides frustrations and disappointments. We believe that there are some positive steps that can be taken to ensure more productive working relationships when consulting with content experts in developing instructional programs.

The creation of successful education or training depends on the availability of technical expertise and content expertise and a supportive working environment in which these people can do their best work. Such an environment is characterized by mutual trust and respect. When anyone working on a project doesn't feel that he or she is respected for their talent or when their opinions are not trusted, the project cannot be as successful as it should be. Often on instructional systems development projects, problems arising from the interactions of the content specialists with the instructional designers and media producers limit both the success of the project and the pleasure derived from working on that project.

So how does a manager work effectively with the staff? First, he or she must listen, and listen carefully, so that the manager can understand others' concerns and points of view. He or she must realize that there are competing interests at work in ISD projects. Just how helpful is your content expert going to be, for example, when he perceives that the purpose of the ISD effort is to develop a CBT training program to replace a lecture course? Is his heart in the effort? Is he likely to be as enthusiastic as you are? Learn to look at the project through the eyes of others.

We hope that as the ISD project progresses, good ideas about how to conduct the education or training will emerge. Be prepared to acknowledge the contribution of others in forming the ideas. Don't hog the project or the ideas. Give credit to your staff and others as often as you can. This reinforcement ought to be done in and of itself, but it will also keep them producing for the project. Be generous with praise and credit; don't grab the limelight yourself.

Summarize and restate often. This is essential to effective communications. Let others know you are paying attention. Try hard to understand what they are saying and don't dismiss their ideas outright. When people have their say and believe that they were listened to, they will be more willing to compromise. Managing an ISD project requires having people reach compromises.

Maintain appropriate distance from your staff and co-workers. As manager you must make decisions and resolve disputes. Thus it is important that you are impartial and that you appear impartial. How effective will you be in resolving a dispute between two content specialists on your staff when one of them is your regular golfing and lunch partner? Even if you can put all aside, how impartial will you be perceived to be by the rest of your staff? It is important that you avoid being placed in a situation in which you respond based on the personalities involved rather than based on your best professional judgment about the quality of the work.

In many large organizations the content specialists assigned to an ISD project will be drawn from another division and will not be members of the ISD staff. Thus the manager does not have direct line authority over the content people on a project. In such situations of divided authority or no authority, frustration can arise. Some content specialists will delight in working with you in developing objectives, creating outlines, case studies, tests, and exercises. They will be creative in coming up with ideas to enhance the project. Other content specialists will view their assignment differently. They will be willing to answer some questions about the content or perhaps to review training materials for accuracy. Perhaps the worse case is a content specialist who simply gives you a few books and reference materials, leaving a very frustrated designer to "dig out" the instructional content from an unfamiliar content area.

Effective managers are consistent. We all experience the ordinary ups and downs in life, but we should not vary widely over time in what we expect from our staff. Your assessment of your staff's work should reflect their work, not your mood. Strive to apply a consistent set of criteria to their products so that they can learn what is expected. In such a manner, your staff likely will give you more consistent work. They will know where they stand

Project Management and Consulting

with you and will realize that the rewards come from doing good work and not from catching you on a good day.

Perhaps the Golden Rule applies as well to managers in instructional systems development projects as anywhere else. Think about what you want from your manager and strive to give this to your staff. Be kind, considerate, and fair. Keep everyone informed about matters that affect them. Listen to their concerns and their problems about the project, and give constructive advice. Involve others.

You should remember that in most large organizations ISD people are seen as a support function. ISD represents a cost center within an organization because it does not generate revenue. In some organizations this might mean that the ISD staff is viewed as second class citizens, since they don't directly provide goods or services to enhance the organization's revenue. This view can hamper ISD projects when a content person's time is divided between training and his or her own functional area. While some large organizations can afford to assign specialists in certain positions to the training department on a full-time basis, many organizations can't. Thus training represents additional responsibilities for people working on other jobs within the organization. The full-time training staff must be aware of job assignments and the overall culture of the organization.

Others within the organization will not share the same knowledge about training or commitment to training that ISD people do. We serve different "gods." What is very important to someone within a training department may be of little value to others within the organization. It is important to be sensitive to this because ISD requires the support of others in an organization. Even when employed as a staff member of an organization, you are essentially a consultant. As a consultant you are serving others. Thus when consulting within an organization, you must become aware of the organization's values and beliefs so that you can be consistent with them. You must explain your work and your approach as if you were selling it, for you often are. Realize that the more communicating and educating about ISD you do, the more your work is likely to be accepted.

In summary, there are some common "people problems" that can be expected on instructional systems development projects due to the diverse nature of the people involved. Be sensitive to these problems; these are as important to the success of an ISD project as the more technical problems.

Chapter 12

Implementing Planned Change

All learning implies a behavioral change of some kind. The acquisition of new knowledge, skills, and attitudes designed to close performance gaps between the current and desired state naturally leads to a change. When a new employee, for example, stops using long, rambling sentences and begins to routinely use shorter ones with more concise wording, that employee has in effect changed his behavior to reflect his organization's desired type of business writing. This individual change will contribute to more effective communication and thus to overall organizational efficiency. The same is true for the head nurse who has learned how to reorganize her day to permit more guidance and closer supervision of nursing students on her floor. Her behavioral change will contribute to overall patient care by producing better trained nurses.

As training designers, we are entrusted with the challenge of planning for this kind of change. We want to implement programs that will ensure the occurrence of new knowledge, skills and attitudes, that they occur in the way they were originally intended, and that they will be lasting.

Most learners will naturally try to improve the way they carry out their jobs, run their homes, and interact with friends, family members, and colleagues—if they can clearly see the value in doing things differently and if there are not too many obstacles associated with the change. However, problems arise in affecting desired change when the new ideas and practices are presented in a way that is hard to understand and/or the environment in which they are expected to exhibit new behaviors is not receptive and supportive. In such cases, the intent of the training is likely to be misinterpreted, require a lengthy period before it is adopted, or simply be rejected or forgotten.

Careful planning can increase the likelihood that a training program will be accepted. It is important for the training designer to think of himself or herself as a change agent whose role is to pay careful attention to the specific needs and expectations of the target audience. As a change agent, the designer needs to ask himself or herself the following questions:

1. What will cause the target audience to value the training and consider it useful?
2. How can I accurately communicate the intent of the training?
3. How can the practice of new behaviors be supported and maintained over time?

The designer must analyze what will cause people to learn and practice new knowledge, skills, or attitudes. He or she must first question whether members of the target audience consider the need for change to be important. If the answer is no, then it may be that the need for change should be represented in a way that has more relevance for the target audience. If that fails, then it may be that the problem is not perceived as a problem at all by the intended recepients. Sometimes the decision-makers who originally requested the training are so far removed from other groups in their community or organization that they may be out of touch with the real needs of their people. Hopefully, a thorough needs assessment would prevent or minimize such misunderstandings.

In evaluating prevailing attitudes, it is also important to determine if these opinions are shared by everyone. Major fragmentation can in time erode the effectiveness of the training. As a rule, time and pressure from other peers will address this problem. Yet if dissenters are either numerous or in a position of power, they may require special attention or a reevaluation of the amount of time it will take to reach all of the training program goals. At the same time, the training designer must look to see if there are any traditions, societal taboos, or political pressures that would inhibit the group's ability to respond to the desired change.

Implementing Planned Change   173

The second question addresses the issue of how new information is received and processed. The training designer must ask if there are cultural or situational differences that will cause learners to perceive things differently. For example, are symbols used that may have a different or offensive meaning for some learners? Are the examples used to illustrate concepts pertinent to the learners' immediate world? Failure to attend to such questions can cause major communication problems, resulting in misunderstandings or the rejection of new information.

It is also important to ask what a training session is expected to be like. Has instruction for the target audience, for example, been the traditional lecture and has most of their learning focused on rote memory? Additionally, does this group have a history of solving problems together, and are they capable of learning from a lesson that may require them to work together to complete a case study exercise? Participants may not take training seriously that does not resemble their notion of what education should look like. The acceptance of any training method is to some extent a function of what the audience expected and the training methods with which they are familiar. In such cases, training designers are counseled to gradually introduce any non-traditional instructional methods that may be used. For example, if the learners have no experience as participants in role-playing or simulation exercises, then any use of these training methods would require careful introduction. Perhaps in such a case the role playing or simulation exercise could be used in the latter parts of a course or in the final courses taken in a training program.

Finally, the training designer, as change agent, must plan for ways to support and maintain newly acquired behaviors. He or she must ask if the society or organization will support and encourage the practice of new knowledge and skills. For example, do supervisors and community leaders have equal beliefs in the value of the training? Are they in positions to help and guide their subordinates? It is a wise idea to include decision makers in the planning of the training program. It gives them a greater sense of ownership and helps ensure that their wishes are being addressed. At the same time, it is important to determine if there are any rules or policies in effect that will discourage long term implementation or reduce the effectiveness of the training. Is the implementation plan so designed that all involved can see some degree of success early on? A sense of accomplishment is important for maintaining enthusiasm for the program. The sooner the program is perceived as successful, the better. Most supporters have difficulty maintaining their belief in the value and the possibility of reaching goals that are perceived as too abstract or too far in the future. Finally, are necessary resources, i.e., funding, equipment, manpower, etc., available to implement the program and maintain it throughout its lifecycle?

Training designers assume many roles as they answer these questions and develop an implementation plan. They may act as a *catalyst* by speeding up or pointing out the need for change. They may play the role of *solution given* by diagnosing and searching for solutions. Often, they may act as *process-helpers* by helping decision-makers, training recipients, and other stakeholders find their own answers to their own problems. Finally, they may act as *resource linkers* because they can use their analysis skills to bring together the varied perspectives and expertise of all the individuals who may be involved in the problem, i.e., decision-makers as in the case of management and staff who will practice new behaviors, clients and other personnel who may benefit from improved services and practices, and outside experts who know the field and can offer an external point of view.

Communication Lines

Training Designers must be able to clearly communicate with all parties involved as they design and implement training programs. Otherwise, they run the risk of gathering invalid and useless information that will make it impossible to truly assess the needs or anticipate the reaction from a target audience. They must, at least figuratively, speak the same language as the beneficiaries and stakeholders. For many years, Third World development efforts were sabotaged because the program designers and sponsors were, for the most part, external experts who viewed progress as an opportunity to emulate their own practices and perspectives. This view was often supported by the national elite whose goals tended to reflect the colonial tutorage of the pre-independence period. Unfortunately, this approach ignored grassroots input and values. Thus the majority of people had difficulty understanding the intent and need for the various training programs. As a result, many early development efforts experienced tremendous difficulties in reaching their goals, and far too often, assisted in further polarizing social, economic and political equity.

Parallel problems with similar lessons to be learned can be documented in industry, the military and government. Regardless of the specific setting, if a training program is to be successfully implemented, then it must be designed to be acceptable to the audience. All parties need to communicate their specific needs and be given an opportunity to react to the training plans. In turn, the training designer should help the different parties communicate with each other to ensure a more shared expression of goals and objectives.

To support communication efforts and to ensure that the instruction is truly communicated in the way it was intended, it is useful to look at communication models. Such models attempt to explain the process by which new ideas and practices are accepted and disseminated. A helpful model to consider is one developed by Katz and Wedell (1978). Their model has seven components:

1. Acceptance
2. Over time
3. Of some specific item (an idea or practice)
4. By individuals, groups, or other adopting units
5. Through specific channels of communication
6. To a social structure (or an organization)
7. To a given system of values or culture.

This model, as many others, describes the process as occurring in the following way. Change agents in the form of training designers, instructors or opinion leaders communicate new knowledge, skills or attitudes through mass media, instructor-led workshops and classes or via self-study materials to members of an organization which is part of a greater community. Over time, the target audience either accepts or rejects expectations for the training program. In turn, if all of these components are considered in planning for change, the likelihood of sustained program acceptance is increased.

Factors That Influence Target Audiences

Cultural Bonds

Anthropologists and educators have long felt that the communication process is culturally influenced. This theory is based on the idea that individuals tend to remember and respond to new information in different ways. This theory also assumes that the process is selective. That is, individuals are programmed to pay attention and remember certain stimuli while ignoring others. Programming of this sort may be based on a kind of internalized set of cognitive codes. It is likely that these codes are shaped by what we know of the world around us which is, in turn, influenced by past and present experiences. It follows that learning is also a selective process and as suggested by Piaget (in Patterson, 1977) and Bruner (1966), one that is based on a filtered version of reality. Most of us will agree that our notion of reality is influenced by the culture in which we live. This is because cultural systems are also based on a type of code. They comprise a behavioral code that influences the actions of its members. With this argument in mind, it is likely that cultural bonds can affect the selection of what an individual perceives and understands.

Increased focus on cultural factors has caused training designers to view training needs as well as the means by which the training is delivered with greater cultural specificity. In addition to more consideration about a given period in a nation's or an organization's history, their internal conditions and external relationships, designers should seek to understand a society's popular culture. The popular culture is the way in which a culture has learned to cope with certain kinds of problems. Eisenstadt (in Schramm and Learner, 1976) defines these culturally-specific means of coping as cultural codes and suggests they will persist even as a culture experiences change.

Shoemaker joins Rogers (1971) in arguing for a greater emphasis on the cultural perspective as a means to help us predict likely communication hang-ups that diffusion campaigns can encounter. They cited the 1952 Apodaca Study of hybrid corn rejection by Mexican-American farmers. It seems that the change agents perceived the corn seed mainly in terms of its higher yield whereas the farmers perceived it as poor food. Another example is the rejection of water-boiling among Peruvian villagers (Wellin, in Rogers, 1973) whose folkways negatively linked boiled water with illness.

Rogers (1973) describes three modes of selectivity in communication. They are selective exposure, selective perception, and selective recall. These selectivity factors are believed to influence one to remember, interpret, and attend to messages in terms of one's existing attitudes, beliefs, and common history (Colby, 1975; Rice, 1980; Rumelhart and Ortony, 1977). These are important considerations for the design of training programs, as it appears that instruction communicated in a manner greatly different from what is common in the existing culture generally requires a longer learning period and often leads to a lack of comprehension.

Varying folkways and customs are not the only important differences to consider in designing training. For anthropologists such as Hall and Reed (1980), cultural bonds are also illustrated in the way groups of people choose to sequence thoughts and events, outwardly express emotion, conceptualize time, determine distance and tactile comfort, associate symbols, and select content. Numerous studies that describe variance in audio and visual perspectives along with patterns in reasoning and social interaction are available in educational, communications, and anthropological literature.

Storytelling research has produced a valuable source of data to study. Stories are useful instruments to study cultures because they describe commonly shared dreams, beliefs, and perceptions of real events. Worth and Adair (1972) have explored story telling in film among Navajo Indians by asking them to use simple filming equipment to create and shoot their own stories. They concluded that it is possible to study a specific culture's "structure of reality" by the way they choose to pattern and perceive images. Colby (1975) uses the term culture grammars to describe differences in narrative storytelling. He suggests that cultures have differing rules for the sequencing of events within a story as well as the development and resolution of conflict between the protagonist and the adversary. In support of these internalized codes, Rice (1980) told Eskimo folk tales to American subjects who were asked to retell them in their own words. The findings suggested that subjects would reorder, delete, or reinterpret narrative information that was culturally foreign. Notable was the Americans' need to assume a happy ending, even if a story had to be modified to provide one.

The examples illustrate, of course, differences in cultures that are greatly different from our own. However, such cultural differences exist not only between different countries and parts of the world, they can exist as well between groups and organizations within the same country. All potential trainees have a cultural alliance of some kind. Alliances may be influenced by religion or national barriers as with the French and Swiss. Differences may also be technological ones similar to those that exist between the First and Third world or on a smaller level between "computer types" and those persons less knowledgeable about computers. Cultural ties may also be linked to an organization or a profession as in the case of the military or a "Big Eight" accounting firm.

We have all heard the expression "you have got to learn the ropes before you can fit in." Contrast, for example, the culture of a major law firm whose clients are mainly large corporations to that of a consulting firm that specializes in relief efforts for the downtrodden. In which one does the hero wear a three piece pin-striped suit? Ask how decisions are made and how problems are resolved. Look at how routine tasks are prioritized and at what makes them bite their nails. In short, all target groups are influenced by environmental factors that represent a cultural bond.

The Attractiveness of the Innovation

Substantial time lags can exist between the introduction and adoption of new knowledge, skills, and attitudes. An issue that often affects the acceptance of a training program is the nature of the innovation itself. Resistance to change is most often associated with the Third World. Yet, even in countries that have a rich history of innovation acceptance, such as the United States,

Implementing Planned Change

change can take a long time. Rogers cites several examples. It took more than 50 years for the U.S. public to adopt the idea of school kindergartens. Now, of course, we have gone beyond kindergarten and have so extended the practice of early care and education that often there are not enough day and infant-care centers to meet the current demand. Another example is the struggle we had in accepting and learning from the modern math approach. It took over six years for this teaching methodology to be adopted in the schools in the 1950s. Finally, a classic story is that of Iowa farmers who required over 14 years before they adopted the use of hybrid seed on a widescale basis.

On the other hand, we can all cite examples of innovation acceptance that occurred almost overnight. Remember the days when simple calculations were done by hand? It seems long ago. Yet, it was only in the mid-seventies that hand-held electronic calculators became widely available and were purchased by the millions. Now shoppers take them to grocery stores and children use them in primary school.

Certain characteristics associated with the innovation itself may, in part, influence adoption rates. We know that innovations need to be perceived as attractive by the target audience. Research findings suggest that innovations are more likely to be seen as attractive when their relative advantages are valued, observable, and involve little risk in trying them out. The key is audience perception. The training designer as change agent must assess the degree to which an innovation's attributes are perceived to be desirable by the beneficiaries. Consider the following characteristics that an innovation must have as suggested by Rogers (1973).

1. Relative Advantage
2. Compatibility
3. Complexity
4. Trialability
5. Observability

Let's now look at each characteristic individually.

Relative Advantage. This is the degree to which economic and social benefits are enhanced. Also included is a sense of increased convenience and personal satisfaction. In short, new ideas and practices will be more readily adopted if they can improve one's lot in life. Take the case of exercising. In the past few years, physical fitness programs have become quite popular. In fact, many companies offer fitness facilities for employees. People believe that exercise will make them healthier and physically more attractive. Thus, economically, it can help lower medical bills and socially it gives people a sense of being a part of a popular movement as well as increasing their attractiveness to the opposite sex. In fact, some sociologists say that health spas have become the new singles clubs of the eighties. People also say that they simply feel better when they are physically fit. The one drawback to the acceptance of exercise programs may be the issue of convenience. Exercising takes time and can hurt if one is not already in shape. It can, also, be somewhat inconvenient to drive to the health club, change and work out after having worked an eight hour day. Many exercise programs can, of course, be done at home.

Despite possible inconvenience, exercise programs will probably maintain their popularity for a while to come because of the perceived relative advantage. Keep in mind that an innovation's attributes do not all have to be perceived to the same degree. In this instance, the relative advantages associated with improved economic benefits, socialization opportunities, and personal satisfaction may override any felt discomforts.

Compatibility. Compatibility is the degree to which an innovation is perceived as being consistent with prevailing values, beliefs, and needs of the persons receiving the innovation. For example, the importance of family planning may be difficult to grasp in a culture where children are traditionally considered to be valuable assets. In such cultures, children may be viewed as additional workers who are expected to contribute to the household income. Numerous children may also be seen as adding to a sort of old age pension plan. The idea is that children will care for their elders in their later years. Family planning programs may have been additionally hampered in societies where birth control techniques are prohibited by religious policies. In such situations, change agents have tried to develop a more compatible approach. The need to reduce population size now emphasizes the health advantages for the mother and her existing children. Training for such programs are retitled to reflect material and child health issues, and the concept of birth spacing is discussed instead of population control.

Complexity. This characteristic is the degree to which an innovation is considered difficult to use. That is, an innovation must be easily understood by most members of the target audience for it to be widely accepted. Computers, for example, terrorize certain individuals. They have been described as too complicated and too difficult to learn how to use. Even the most rational have expressed fear of computers blowing up, or worse, losing all their data. However, with the advent of widespread office automation, many individuals have been forced to overcome their anxieties. Training designers have been busily developing user friendly training courses designed to promote the computer's worth by making them easier to understand in a way that bolsters confidence. This is a classic case where the relative advantages of the innovation will shortly outweigh the perceived difficulties.

Trialability. Trialability is the degree to which new innovations may be introduced on an experimental basis. It thus appears that new ideas and practices are perceived as less threatening and better accepted when they can be tried out in stages. Consider the case of language training. There are many ways, of course, to learn a new language. The best method is probably to totally immerse oneself by living in a country for a period of time where the language is spoken. Total immersion, however, represents a major investment in time and money. What if the individual, for example, doesn't like the sound of the language, has difficulty learning it, and feels uncomfortable with the culture? In such cases a major investment may be translated into a major loss. Most individuals are hesitant to take such risks and would prefer an opportunity to see if they are likely to like learning and using the language. Thus, it may be preferable to begin with a few night courses followed by some trips abroad before one packs up home and hearth.

Observability. This characteristic refers to the degree to which results are visible to others. The easier it is to see results, the more likely the innovation will be used. Let's suppose, for example, that as part of a training program on animal care, farmers have been encouraged to use a new kind of feed for their hogs. Let's also suppose that they were encouraged to do so because specialists in animal husbandry report that the animals tend to like this brand better than others. If the price is the same and requires no additional effort on the farmers' part, they will probably make the change. However, if added effort is required, then they will need a more concrete reason to buy the new feed, such as, the sale of higher quality pork yielding higher market prices. Associated with this point is the amount of time it takes to see results. Farmers will be encouraged additionally if they can see results in a year rather than four or five years.

While all these variables are likely to influence adoption rates for a given innovation, it is unlikely that all of these attributes will be perceived to an equal degree by the target audience. In turn, a variable that is perceived to be highly attractive may override and counter the negative effect of an attribute that is not perceived as strongly. This aspect was especially true in our example on the use of computers. The increased convenience, speed and productivity associated with computers may override fears that they may be too complex to use.

Differences in Adoption Rates

As discussed in the last section, the acceptance of new knowledge, skills and practices can be affected by a number of different variables. It is also important to remember that innovations are seldom accepted by all members of a target audience at the same time. This phenomenon is true within a given culture as well as in international education where several different cultures may be involved. It is also true within a corporation. The issue is not the overall cultural appropriateness of the design. The phenomenon also exists and must be anticipated in implementing programs that are appropriately designed to reflect the culturally influenced practices and more of the participants. The issue concerns adoption rates that are beyond the control of the specific design itself. Knowledge of how adoption rates can vary is important in planning the piloting of programs and implementing them on a wide-scale basis. This section will describe the nature of those differences, while the final portion of this chapter will suggest how to best address variance in developing implementation plans.

Implementing Planned Change

Rogers and Shoemaker (1971) have categorized individual adoption styles in classes in an attempt to describe individual differences within a cultural system. The following model is based on their writing.
1. The Innovator
2. The Early Adopter
3. The Majority Adopter
4. The Laggard

The Innovator. The innovator is basically a risk taker who lives on the fringe of the status quo. He may be perceived as a fanatic or unrealistic dreamer. The innovator is the first to lead protest parades and sign petitions. The innovator really doesn't care what other people think of him and thus may practice a lifestyle that is quite different from the norm. Innovators often introduce new ideas and practices into a society. However, they rarely influence large numbers unless the benefits are clearly obvious.

The Early Adopter. This individual is perceived as a conservative pace setter because he represents a balance between traditional and modern thinking. Most have had an opportunity to have positive experiences outside of their home environment. They are thus sufficiently cosmopolitian to recognize the value of new ideas and practices and sufficiently traditional to be believed by the majority. In this sense, early adopters often act as opinion leaders and tend to number among a society's leading citizens or a company's most valued employees.

Majority Adopter. This category of individuals reflects general public opinion. As a group, they value tradition and tend to maintain the status quo. They are the ones who uphold the practices of those who came before them. While they recognize that change is sometimes healthy to maintain the vitality of a society or the effectiveness of a corporation, they do not actively seek it. They will thus take their time evaluating an innovation before accepting or rejecting it. They are not risk takers. They will be comforted in the knowledge that an innovation represents continuity with the past. They will, in particular, be more open to a gradual sense of change than one that appears abrupt, causing a major break with historical practices.

Laggard. This group is the opposite of the innovator. In fact, they may be resistant to change simply because it is a change. These individuals have a strong desire to keep things as they are and often may even have a vested interest in doing so. Training may have little effect on this group. As a rule this group changes only as a result of group pressure. In short, the group will be the last and hardest to reach. Given their stubborn and negative attitudes, it may not be cost-effective to try to reach them through any special efforts.

The four categories thus represent a range in adoption rates from the very willing to the very unwilling. Those willing to try new ideas and practices tend to have the following traits (Beal *et al.*, 1957; Beal and Rogers, 1960):

1. Higher education.
2. Cosmopolitalism; a broader world view.
3. Rational rather than emotionally-based behavior.
4. High value on science.
5. Less interested in maintaining the status quo.

Although the innovator may be more willing to change, training designers, acting as change agents, generally have better success working with early adopters to introduce training programs. This group is both open to new ideas and better able to influence others than the innovators. Innovators are generally not perceived as sufficiently traditional to gain the respect of the remaining target audience. For example, if an innovator decided to stay home and replace his wife as the person most responsible for child care and for the home, he might be perceived as very unusual or even threatening to other men. On the other hand, if an early adopter practiced the same behavior, he might be seen as a trend setter.

Developing an Implementation Plan

Training programs obviously must take into consideration a number of issues, as we have seen in the previous section. It takes time and a well developed plan to ensure a reasonable degree of success. Time is needed to correct unanticipated problems and, as discussed earlier, time is needed for all members of the target

Implementing Planned Change

audience to accept and value the intentions of the training. Implementing a graduated approach of the training program is generally the best strategy. This kind of strategy suggests offering the training in stages rather than trying to reach all of the target population at once.

A staged approach implies that all individuals from the target audience can be placed into different groups. Each group will, in turn, be assigned a different time to receive the training. The rationale for group placement may vary depending on the nature of the program and its environment. Common variables that may be used are associated with geographical or departmental differences. Group selection may also be based on number of years with the organization or the likelihood that a given set of individuals may be more resistant to change than is expected as a result of the training. Schedules for conducting the training for each of the different groups may also vary. Some implementation plans will schedule stages back-to-back. For example, once group A completes the course or series of courses designed for their target population, group B will immediately begin their training. In some instances there may be a lapse of weeks or months between stages. On the other hand, in some programs, a group may begin their training before the previous one has completed theirs.

An additional characteristic associated with implementing a program in stages is that it permits an opportunity to formatively evaluate and revise the program before it is conducted on a widespread basis. This can save time and money and ensures that any mistakes are committed on a small scale which can easily be corrected. Another advantage of this approach is that the target audience may perceive less risk when a program is introduced on an experimental basis. This is especially important to majority adopters whose readiness for change is enhanced by advance program support and demonstrable results. It is thus likely that a staged approach can result in better designed programs, quicker adoption rates, and a more efficient use of time, personnel and money.

Piloting the Program

The first stage of any program is commonly called the pilot. Pilots constitute the first time that the training is actually conducted. In turn, they represent the first chance for the designers to assess the feasibility and appropriateness of chosen strategies. Its results are critical for both training designers and decision-makers who must justify their initial and continued support of the program. In turn, decision-makers need to see pilots that result in a reasonable degree of success. To enhance a pilot's potential for success, the following variables should be considered in selecting a pilot.

1. All those involved should be eager to participate.
2. Greater than average difficulty should be expected in implementing the pilot.
3. Pilots should be logistically accessible to decision-makers.
4. Size and scope should be small enough for in-depth experiences.

All those involved with a pilot program should be willing and eager to participate. This includes community or departmental leaders, the participants, training personnel, and individuals who represent top decision-makers. While a uniformly enthusiastic group may not be the most representative they will ensure the patience and endurance required to work with an "untried" program.

Greater than average difficulties should be expected in implementing the pilot. Again, a representative set of resources and constraints may not be the key. Pilot programs that overcome weak resources are better able to counter criticism. This point is especially valid if a pilot is perceived as unrepresentative because its participants were thought to be more receptive than the average target audience.

It is helpful if pilot programs are logistically accessible to decision-makers. It is important for these individuals to be able to watch training in progress without, of course, disturbing the participants and for them to observe the results. The act of being physically on site will make the program appear more real to decision makers. Most importantly, however, is the fact that successful results are more believable if they can be seen.

Finally, the size and scope of the program should be small enough to ensure in-depth experiences. Smaller pilot programs generally yield richer data for formative evaluation purposes.

Formative evaluation will, of course continue as the program continues. However, pilots mark the first enactment of the program. For this reason they generally result in the largest revision effort.

Once a pilot is selected, it is important to expect and judge success in realistic terms. This is best done in small doses and is especially important in the first year when the chance for success is the most risky. It may be more feasible and easier to measure success in terms of statistically significant differences rather than quantifiable percentages. Statistical differences are useful because they generally do not have to be as dramatic or visually as large. Planners are simply saying that a change has occurred that is important enough to support our continued efforts. Additionally, planners are not locked into producing a given set of numbers. For example, it may be unrealistic, not to mention risky, to expect a 10 or 20 percent increase in productivity or profit. Instead it may be more realistic to simply anticipate a difference that is significantly different from what occurred in the past.

Continued Planned Considerations

The training designer must continue to gather data once the program continues past the pilot stage. In fact, most designers and decision-makers will support an on-going assessment plan throughout the life of the program. In addition to the assessment of macro (program and curriculum) and micro (course and lesson) level goals and objectives, the following data are needed to ensure an effective on-going implementation approach.

1. Training needs for personnel assigned to implement the program.
2. Potential for incorporating additional personnel into the project.
3. Adequacy of organizational linkages and support.
4. Use of resources and strategies to overcome constraints
5. Success criteria required to initiate each stage.

It is first important to consider the training needs of personnel assigned to implement the program. Such personnel may include, for example, administrators, trainers, evaluators, logistical coordinators, and technical experts. The training designer must assess their current capabilities to carry out their assigned roles. This issue is especially critical if personnel are new or were not involved in the initial needs assessment and plans for the training design. It is also important to remember that personnel assigned to implement the program may differ with each stage of the program. Thus, the need for assessment and training should be a part of each stage's preparation.

Second, the possible need for additional personnel should not be overlooked. Perhaps all roles were not anticipated in the initial plan or perhaps situational differences may require new and different personnel as the program continues. It is also important to encourage the beneficiaries to take on more responsibility as the program continues. This "transfer of power" permits the client organization or group more responsibility and "ownership" in their program, which results in less reliance on external training design assistance. In either case, specific and general orientation needs should be considered with the start-up of each stage.

It is also important to determine if organizational linkages and support are adequate. More specifically, are there assigned organizational representatives involved that can ensure the organization's commitment? Are these individuals communicative, open-minded, and accessible? Too many programs have experienced unnecessary difficulty because the assigned representatives responsible for providing organizational assistance and other types of resources did not have the time, interest, or power to carry out these tasks. In such instances, it is wise to offer additional orientation that may include program benefits, time and resource management to enhance their interest and ability to provide required organizational assistance. If that tactic fails, it is probably best to request another individual to help with implementation plans.

Another question to ask concerns resources and constraints. Is maximal use being made of available resources, and have all alternatives been considered for dealing with the constraints? Training designers should be alert to the possibility of situational changes which result in changing resources and constraints. This issue should be reevaluated before beginning each stage of the program. Naturally, this is where firm organizational assistance can make a difference.

Finally, a graduated implementation approach requires a continued assessment of criteria to use to initiate the next stage of the program. These questions might include: Where will the next stage occur? How do we recruit the support of early adopters? What variable will be used to determine success? What difficulties must be anticipated and overcome?

Finally, it can not be emphasized enough that a staged implementation plan encourages the beneficiaries to become increasingly more active in carrying out their own program. Be it country X or corporation X, the locals may initially welcome outside assistance but will continue to maintain an underlying suspicion of external involvement. A sense of ownership makes the training seem less foreign and thus encourages a sense of greater relevancy and value. Ownership also implies that the need for change comes from within and is thus not forced upon them. It also leads to less dependence on outsiders to manage and revise the training as needed.

Summary

Careful planning is needed to implement a program that accurately communicates the intent and value of training. Problems can arise when the training's intent is misinterpreted, rejected, or requires a lengthy period before it is adopted. Training designers, as change agents, require clear lines of communication to assess and anticipate an audience's reaction to a new program. The probability of program acceptance can be strengthened by anticipating influences associated with culture, the innovation itself, variance in adoption rates, and the need for a staged implementation.

Selectivity factors associated with one's culture may affect how new information is understood and remembered. Training that is designed and communicated differently from the popular culture generally requires a longer learning period and leads to a lack of comprehension or misunderstanding.

The acceptance of new ideas and practices can also be influenced by the degree to which certain attributes are perceived to be true about an innovation. Innovations tend to be more readily accepted if their relative advantages are valued, observable, and if they can be easily used on an experimental basis.

Adoption rates may also be affected by individual differences in the target population. Beal uses four classes to rank-order differences. He associates an openness for change with cosmopolitanism, rationalism, risk-taking, and a high-value on education and science.

Finally, successful implementation is generally associated with a graduated approach. This method permits a chance to formatively evaluate the program, presents less risk for decision-makers and participants, and promotes advance support and a readiness for change through demonstrable results. This plan also demands a more active involvement by the beneficiaries in the ownership of the program. A sense of ownership is needed to ensure success because it encourages the target population to perceive the training as more relevant and therefore more valuable.

Chapter 13

Evaluation in Instructional Systems Development

An inherent aspect of any instructional systems development effort is on-going evaluation. Most ISD models include evaluation and revision as a necessary step in the process. In fact, whether you are "doing ISD" may be questioned if there is no evaluation. Evaluation is essential in an ISD project; there is widespread agreement on this. There is some question, however, regarding how to go about conducting an evaluation. There are different approaches to evaluation and different models for designing evaluation studies. The intent of this chapter is to focus on different conceptions of evaluation. Rather than giving the reader a prescription to follow for evaluating education/training programs, we will offer various evaluation approaches and models from which to select, based on your own set of circumstances.

Purposes of Evaluation

Since there seems to be agreement that evaluation is an important step in ISD, it is interesting to examine why evaluation is thought to be so essential—to examine what purpose it serves. As previously stated, the intent of instructional systems development is to improve the performance of educational and training programs. This improvement might take the form of enhancing the effectiveness or efficiency of the program. Since the goal of instructional systems development is this improvement, we must know whether the effectiveness or efficiency of the educational or training program was improved. In short, it is only through evaluation that we can determine whether the educational/training program is doing what we had intended (Maanen, 1979).

Further, it is primarily through evaluation that we can discover what changes to make in educational/training programs to improve their performance. Evaluation results are also used to determine whether individual learners have adequately met the objectives and are thus ready to advance to the next level in the education and training program or to be placed in the job for which they were being trained.

There are different orientations regarding the purposes of evaluation. Within an ISD setting we believe that the main purpose is to determine whether the program is working and to provide information for revision. This approach to evaluation is most closely related to the models of Tyler (1969), Provus (1971), and Stufflebeam *et al.* (1971). The common thread through these evaluation models is the collection of data about learners' performance with regard to the intended goals. The data then are used as the basis to judge the worth of the program and to guide any modification of the program. Other evaluation models rely more heavily on professional judgment rather than student data for making decisions about a program. Some models seek to avoid measuring the attainment of specific goals but rather seek to document what the evaluator observes happening. There are many competing approaches to evaluation with rather strong differences regarding the appropriate purpose for evaluation and how evaluations should be conducted.

Evaluation and measurement. At this point it seems desirable to distinguish between evaluation and measurement. Regardless of the differences among various approaches to evaluation, there is an agreement that evaluation implies some judgment of worth or value. As you may expect, different evaluation approaches use different criteria for judging worth or value and have different ways of going about making these judgments. All evaluation approaches, however, seek to make some judgments about the worth or value of a program. Measurement, on the other hand, is that process of systematically assigning some number to an entity. We measure something when we represent some aspect of it numerically. We measure our height by determining how many feet tall we are, our weight by determining how many pounds we weigh. Good evaluation efforts require good measurement. There are many types of data that might be collected in an evaluation. Typically a distinction is made between data that reflect mastery of objectives: "hard data"; and data that reflects the learners' judgment of the training's success or their pleasure with the training they received: "soft data." In these examples we are assigning a number to some trait according to a set of rules. When we start making a judgment about how appropriate our weight is, we are in the realm of evaluation. Measurement is the process of determining the status of certain variables, say mathematical achievement, following an educational program. Evaluation is the process of determining the value or meaning of these outcomes. Measurement seems to be forever involved in evaluation but, as you see, they are separate processes. An insurance company may have established a training program to acquaint its agents with new methods for determining the costs of policies with different coverage. To determine whether this training program was effective, they would have to determine how much the participants learned about calculating costs (measurement) and decide whether the training program was worth the effort (evaluation). While it is typical to find measurement and evaluation associated, it is necessary to separate them for discussion purposes. Much has been written about measurement, from theoretical works to practical guidelines, and there is much to be learned about measurement. Yet measurement seems a technical specialty that does not have to be mastered by persons with responsibility for developing and managing educational/training programs. Certainly measurement expertise is important in evaluating educational/training programs. It just seems that managers and directors of training programs should focus more on the evaluation aspects than the technical aspects of measurement. Furthermore, since the emphasis in this book is on designing and planning educational/training programs, this chapter on evaluation should address evaluation concerns at the program level, not the level of the individual learner. The assessment of individual learners is best discussed in the literature on tests and measurements. Thus the focus in this chapter will be on program evaluation for managers of large-scale educational/training programs.

Uses of Evaluation Results

The data collected during an evaluation serve many purposes in instructional systems development. They are used to (1) determine the success of each person participating in the program for remediation purposes or when certification is required, (2) assess the effectiveness of the instructional procedures and materials used, (3) decide if revisions in the educational/training program are necessary and, if so, where they should be made, and (4) determine the usefulness of the training once the trainee has returned to the job. Evaluation data are essential to the ISD process, since it is through evaluation that we determine the effectiveness of the educational/training program. In an ISD approach to education and training the instruction is designed to achieve some purpose, to reduce some documented need. The instructional program and any aspect of it is subject to revision until it causes the program's needs to be met. Evaluation data are essential to this process. With good evaluation data, programs can be revised until they are effective; without evaluation data or with faulty evaluation data such revisions are not possible.

Evaluation serves a navigation function within the overall ISD process. It is through appropriately conducted evaluations that ISD improves education and training programs over time. Evaluation helps program managers meet their training goals in a timely manner. Cost-benefit assessments of education and training programs require careful evaluations to determine the benefits derived from the training. In fact, evaluation data are often used to support funding decisions because these data offer "proof" of the effectiveness of the training programs. Top level management can review evaluation data to determine the success of their investment in training.

Approaches to Evaluation

There are four primary approaches to evaluation that differ in their beliefs about the intent of evaluation and their focus. Evaluation has been viewed as:

(1) the collecting of information about education/training programs for the purpose of decision making: the DECISION MAKING MODEL;

(2) forming of professional judgments about the processes used within the educational/training programs: the ACCREDITATION MODEL;

(3) determining whether prestated goals of educational/training programs were met; the GOAL BASED MODEL;

(4) uncovering and documenting what outcomes were occurring in educational/training programs without regard to whether they were the intended program goals: the GOAL FREE MODEL;

Decision making model. In the decision making model the intent is to collect relevant evaluation data for the purpose of informing decision makers regarding the effectiveness of their programs. Data about several aspects of an educational/training program are collected and analyzed to provide a wide range of information to decision makers. Information is collected in four areas: (1) the context of the program, (2) the inputs to the program, (3) the processes used in the program, and (4) the products emerging from the program (Stufflebeam *et al.*, 1971). Since the decision making model collects such a wide range of evaluative information, it is the most comprehensive evaluation model. In fact, it may be considered as four evaluations: one for context, one for inputs, one for process, and one for products.

The **context evaluation** focuses upon the situation surrounding the educational/training program. This part of the total evaluation provides the decision makers with data about the environment in which they are operating. The basis for developing the goals and objectives of the educational/training program can be established by examining the context or external environment of the program. Thus the context evaluation can serve to aid the planning process by identifying unmet needs and problems in the environment. In this sense, context evaluation parallels needs assessment.

The **input evaluation** seeks to determine what resources are available and how decision makers might use these resources to meet program goals and objectives. In order to do this the input evaluation examines the capabilities of the agency or organization, various strategies for meeting program goals and objectives, and optional designs for implementing a selected strategy (Stufflebeam et al., 1971).

The **process evaluation** occurs after the program is underway and monitors progress in an ongoing fashion. The process evaluation also tracks what is taking place in the educational/training program. The process evaluation provides decision makers with timely information about the program's operation and any problems encountered. In essence, the process evaluation serves as a navigational system to the day-to-day managers of the program.

The **product evaluation** seeks to assess the outcomes or products of the educational/training program. Product evaluation typically includes measures of the attainment of objectives. Process evaluation determines the extent to which the procedures in the program are operating as intended; product evaluation determines the extent to which the objectives of the program are being met.

This decision making model is comprehensive in that it requires the collection of information from a variety of sources about many aspects of a program's operation. It reflects a systems approach to evaluation, touching on almost every aspect of a program's operation from initial planning, daily management, and assessment of accomplishments. Thus this decision making model is consistent with an overall ISD approach than other models.

Accreditation model. The accreditation model focuses upon the use of professional judgment to assess the worth of an educational/training program, particularly the processes used in the program. Rather than seeking data from participants on program outcomes, the accreditation model relies on the opinion of knowledgeable observers to assess the effectiveness of the educational/ training program. Typically persons who are familiar with the type of educational/training program being evaluated are called in to conduct the evaluation. Through observation, review of documents, and select interviews, the evaluators collect information about the program's operation. Finally, the evaluators make a set of recommendations about the program's operation.

In the accreditation model the focus is usually on the educational or training processes used within the program. The people doing the evaluation determine the extent to which they believe these processes are acceptable. Hence the need for experts to serve as evaluators. The evaluators must determine whether, in their judgment, the program is operating effectively and how to improve it. The basis for these decisions is the evaluators' professional judgments regarding what is happening during the training.

As you may expect, the accreditation model can be implemented much faster and will provide information sooner than the decision making model. The accreditation model does not require the establishment of an evaluation capacity within your organization, nor does it require elaborate data collection. Rather, the accreditation model requires that you assemble a few experts to serve as evaluators, allow them several days to interview your staff and observe your programs, and have them report their findings to you. Because it relies so heavily upon the judgment of just a few persons, the accreditation model has questionable objectivity and empirical basis. Usually the evaluation team would meet prior to their observations and agree on what they are going to look for and the criteria they will use. While you get evaluation results quickly, the results are somewhat a function of the evaluators' own biases. Certainly you might get some helpful ideas about improving your educational/training program, but you must be cautious about how to use the information you receive.

Goal based model. The goal based evaluation model seeks to ascertain the extent to which the educational/training program is meeting its stated goals. The purpose of a goal based model is to determine the degree of goal attainment of the persons being trained. In a goal based model the performance of the trainees is compared with the stated objectives before and after the training to determine the program's success. Indeed, the pre-post measure of performance of the trainees or students is the main characteristic of the goal based model.

The goal based model does provide good information about the attainment of the project's goals and objectives but is criticized for being narrowly focused. Goal based models would not provide information about the processes followed in the educational/ training program, nor would it be likely to identify unintended outcomes that may arise from the program. Thus, the results of a goal based evaluation are limited to the assessment of trainee performance vis-a-vis the stated goals and objectives. Since the educational/training program exists to help people reach these goals, a goal based evaluation model may seem appropriate. Such a model, however, will not help us identify what part of our educational/training program might not be working as we wish. A goal based model provides little information about the processes used in the program. It may be that an educational/training program was meeting its objectives in terms of what the trainees were learning, yet there were some unexpected negative outcomes that accompanied this. A goal based evaluation would indicate that the goals and objectives were being met by the program but would fail to identify the negative consequences. For example, all of the trainees could be meeting the objectives of the program, but a large portion of these trainees left their jobs shortly after the training. Perhaps the trainees met the program's objectives but their actual performance back on their jobs was worse than before the training! Unfortunately, these unintended things can and do happen in educational/training programs; a goal based evaluation model will often fail to recognize them.

Goal free model. Another approach to evaluation is a goal free evaluation model. As the name implies, in goal free evaluation the evaluator does not approach the task by attempting to measure progress towards predetermined goals. Rather, the evaluator seeks to document what *effects* an educational/training program is actually having. That is, the evaluator does not start by looking at a program's goals. He or she starts by trying to determine what is *accomplished* by the program. By taking this approach, the evaluator is more likely to identify any unintended outcomes or consequences that may be resulting from the educational/training program. The evaluator may look narrowly at the training and trainees or he or she may look beyond the training setting, following the trainees on their jobs. Of course, this evaluation of the impact of training on one's job performance may occur in other evaluation models as well.

In the goal free evaluation approach the evaluator begins by examining what is being accomplished in an educational/training program. The task is to describe and document the outcomes of the educational/training program and to do so before examining the stated goals of the program. The assumption is that knowing a program's goals prior to evaluating the program might act as a set of blinders that obscure outcomes other than specific goal attainment. Thus, the evaluator first conducts the evaluation and only then compares the evaluation results with the program's goals. Proponents of goal-free evaluation believe such an approach will come closer to capturing the reality of a program's accomplishments.

Opponents of goal free evaluation think that this type of evaluation might miss assessing a program's goal attainment. Yet, they might agree that a goal free evaluation could be more sensitive to unintended outcomes. However, they hold that a goal-free evaluation might miss the mark by not providing data on all of a program's goals. It might not have occurred to the evaluator to assess certain outcomes that may turn out to be the main objective of the educational/training program.

Summary. In summary, there are four different approaches to evaluation that might be considered when planning the evaluation of an educational/training program. Each approach has advantages and disadvantages; there are trade-offs to be made.

Evaluation of Education/Training Outcomes

Any of the different approaches to evaluation can be used to assess and evaluate different outcomes from education/training programs. What is to be evaluated depends on the intent of the training as well as constraints on the evaluation. There are many aspects of a program that can be evaluated and many different variables on which data can be collected. For ease of discussion, we will classify the possible outcomes of education/training programs that could be evaluated into four different groups based on the work of Kirkpatrick (1987).

One option is to evaluate the trainees' *reaction* to the education/training. You could measure the degree to which the trainees were satisfied with the education/training. You could have the trainees indicate their opinions about the value of the education/training. They could be asked to rate the success of the education/training.

In each case, the trainee or student is providing his or her reaction to the training.

A second type of outcome from education/training programs that is frequently evaluated is the degree of learning acquired—the trainees' *achievement*. Evaluation of the learners' achievement is usually what is first thought of when people plan an evaluation of an education/training program. This is only natural since education/training programs are designed to lead to mastery of the instructional content. The achievement of the trainees is perhaps the most important outcome sought in the training and thus it is often the basis for the evaluation of the success of the training.

A third type of outcome from education/training programs that might be evaluated is the resulting *job performance* of the people following training. An often stated rationale for training is that it leads to improved job performance. If training is seen as a means to enhance job performance, which is usually the case in large organizations, then evaluation of the impact of training on job performance seems warranted.

A fourth education/training outcome that could be evaluated is the *impact* of the training on the organization. This is a broader outcome than job performance and examines the impact of training on the organization's accomplishment of its goals. It measures the contribution of training to the organization.

It is difficult to separate out the effects of training on job performance or on the organizational impact from other factors that influence job performance. The further we get from the training, the more difficult it is to say with certainty that the training caused any effect we may find in the evaluation. Thus, we can determine with more certainty that the trainee's reaction and achievement were a function of the training rather than outside factors. It is more difficult to conclude that changes in job performance or impact on the organization were caused by training.

The following sections in this chapter deal with the concern for the quality of the evaluation data that are collected—regardless of which evaluation outcome is sought. Care in planning and carrying out data collection will increase the faith we can place in the evaluation results.

Data Collection

Regardless of the evaluation approach used there are some concerns about the design of the data collection that must be addressed. Since you will use the evaluative data for the purpose of making judgments about the training, it is important that the design for collecting data be as "clean" as possible. Also, since the intent is to assess the educational/training program, the logic of the design should be such that it is possible to infer that any changes were a result of the program. In essence, you must consider different designs for the evaluation study. Perhaps the best way to approach this situation is to examine the three categories of models for research specified by Campbell and Stanley (1966). They grouped models into pre-experimental, experimental, and quasi-experimental based on certain characteristics of the models.

Pre-experimental designs. A pre-experimental model is one that fails to adhere to the tenets of basic experimental design. A common example would be an evaluation study that consisted of only one group that participated in some instructional program and then was assessed for the result. Let's suppose that the results from the assessment were positive. Can we safely conclude that the program was a success? On the surface, the answer would be yes. The group that participated in the educational/training program had high scores following the instruction. The issue is, can we attribute their scores to the *instruction*? Might their scores have been even higher had there been no instruction? Is it conceivable that their scores were higher before the educational/training program started than at the end of the program? If we have only one group that participates in the educational/training program, then we can't reliably know the answers to these questions. We could, of course, test the participants both before and after the training program. We would then know whether the scores following the program were higher than those prior to the program. That would seem an improvement since it would reduce one source of possible error in our evaluation. Even so, we still have the problem of attributing the results to the educational/training program. It is one thing to determine that the performance of a group of trainees rose; it is another thing to demonstrate that the cause of the rise was the educational/training program. Yes, the scores increased and, yes, the trainees participated in the educational/training program. Is that conclusive evidence that the program caused the increase? It may provide a basis for arguing that the program was successful, but it hardly provides *proof* for that claim. There are many other possible sources of influence that could have produced the increased performance. Perhaps the improved scores just reflect the trainees' normal growth over time and not the educational/training program itself. That is, it is possible that the improvements might have happened anyway, without the educational/training program. Maybe some of the trainees' managers began taking a more active role in employee training, and the improvements resulted from that attention—not the educational/training program. Maybe the persons being trained were those who had an interest in learning more and would have done so absent any educational/training program. You might assemble another group of persons that don't participate in the educational/training program. The performance of the group receiving the training could then be compared with the group who do not receive any additional training. In such a manner you would have a basis for comparing the performance and possibly eliminating some of these competing explanations for improved performance. If the managers were doing more instructing directly on the job, then this would be reflected in both groups. Since both groups would have received any benefits from attention from managers and if the groups' performance differs, you eliminate the possibility that these differences were due to something the managers did rather than the educational/training program. Having a comparison group helps when interpreting the evaluation results and trying to determine the causes of any differences that might be found. However, unless people are assigned to each group on a random basis, there is a problem in assuming that the groups are equivalent. The case is often that the groups are not equivalent. Employees at one site might differ from employees at another. Using site A as the training location and site B as a comparison site may create problems. Perhaps the employees at site B are older, more experienced, more motivated, have more prior knowledge, etc. While this comparison group design begins to approach a quality design, it falls short because the groups might not be equivalent.

So far we have examined three common models for the design of evaluation studies. Each model suffers from serious flaws that may lead us to form incorrect conclusions. The single group post-test only, single group pre-test post-test, and non-equivalent comparison group models are all considered pre-experimental models since they fail to conform to the basic rudiments of experimental science. Because of the seriousness of the flaws in these models many persons recommend not using such models. The use of these models in evaluation studies may lead to erroneous conclusions and poor decisions about the educational/training program.

Experimental designs. True experimental models feature a comparison group that is thought to be equivalent because each person is assigned to a group by a random process. Of course, it is hard to "know" in some absolute sense that the two groups are in fact equivalent. Random assignment, however, gives us a greater probability that the groups are equivalent in every aspect except that one group participates in the educational/training program and the other group does not. Thus, any differences found between the groups can reasonably be attributed to the program. While this is the most desirable model for evaluations, it is somewhat rare. The problem with implementing a true experimental model in an evaluation study is the difficulty with random assignment and the problem of withholding the instruction from the comparison group. It is unusual for a company to hire, say, a hundred new employees then randomly select fifty for training and withhold the training from the other fifty so that their evaluation might be more scientifically valid. This may be desirable but often times difficult. Actually a better design would require four groups so that you could determine the pre-post gains from the educational/training program and so that you could estimate what effects the pre-test itself had on performance. It has been documented that people learn something in the process of taking tests and that a test before some instruction makes people more sensitive to the instruction. Thus if we pre-tested a group, had them participate in an instructional program, then post-tested them, we wouldn't be able to determine what effects the pre-test had on their subsequent learning. Their post-test scores are a function of the pre-test and the instruction, and we would not be able to determine the relative contribution of each. This becomes important in an evaluation since we seek to determine the effectiveness of the training program. We don't know what would have happened if the group had received only the training program, not the pre-test. The way around this dilemma is to randomly assign people to four groups: one that has a pre-test and instruction, one that has instruction only, one that has a pre-test and no instruction, and finally one that has neither pre-test nor instruction. Then you can compare the post-test performance among these groups and estimate the effects arising from the instruction and from the pre-testing. This design presents even more practical problems. Now there is a need for four groups which requires more people. The assignment must be at random. The instruction must be withheld from two groups. Again we find the requirements of acceptable evaluation designs competing with the practical needs of any organization. The more "scientific" evaluation models are simply less practical when applied in the real world. The fallback position is the use of what are called quasi-experimental designs.

Quasi-experimental designs. The quasi-experimental designs differ from the true experimental designs in that they don't have random assignment to treatment and comparison groups. One commonly used quasi-experimental model has two groups that are not assigned at random but that are constituted to be as close as is feasible. While this is less acceptable than the true experimental model, it is more practical. It is possible to take employees at two different sites, provide training at one site and not the other, and compare the results. Perhaps we could examine other information about the employees at both sites to determine how close the groups are in regard to certain factors that might relate to their performance. That is, we could check to determine if the two groups were roughly equal in terms of prior education, years of experience, and the like. If they appeared equal then we would have more confidence that any differences noted in the evaluation were due to the training program, not some pre-existing differences.

Another possible quasi-experimental model uses multiple measures on one group over time. In this approach the evaluator takes several performance measures before the educational/training program begins and takes several measures after its completion and examines the data for trends. If all the measures taken prior to the program were at a consistent level and that level was somewhat below the consistent level of measures taken after the program, there is evidence that the program caused the change. If just two measures were taken, one before and one after the program, then there is more room for doubt that any differences found are systematic or result from the educational/training program. On the other hand, if there are many measures taken and there is a consistent pattern of higher performance following participation in the program, then the evidence that the program was successful is more compelling.

Summary

Evaluation plays a major role in instructional systems development programs by helping determine the effectiveness of the training. There are four approaches to evaluation: the decision making model, the accreditation model, the goal based model, and the goal free model. Care must be taken in conducting evaluations so that you will be reasonably able to ascribe the results to the training program. The role of pre-experimental, experimental, and quasi-experimental designs were discussed as they relate to evaluations. Program managers are probably most interested in evaluating (1) learner mastery for certification purposes, (2) program impact to determine whether it was worth the time and money, and (3) the design of the training program itself. Evaluating the impact of a program is the most difficult and perhaps the most important in the eyes of management at all levels.

Chapter 14

Future Directions for Instructional Systems Development

Instructional systems development (ISD) procedures have had an impact on the way education and training programs are planned, developed and delivered. Perhaps the landmark event in ISD was the Interservice Procedures for Instructional Systems Development project in 1975. This document offered a model for how to use ISD in the military that became official doctrine. The development of training programs in business and industrial settings was also influenced by the ISD model from this project. In essence, this ISD model set *de facto* standards for the development of education and training programs. Because of the demonstrated effectiveness of ISD procedures, we should expect to see wider use of the procedures in the future. To date most ISD efforts have proceeded in a remarkably similar fashion, following a set of steps or procedures that approximate the ISD model used in the military. While ISD efforts based on this model have met with success over the past decade, it seems time to "revisit" such ISD models to determine whether they offer proper guidance to education and training developers in today's climate. Do the ISD models reflect the current state of the art regarding education and training, or have there been improvements in the years since the ISD model was designed that should be incorporated into instructional systems development efforts?

Many research projects have been conducted in a variety of fields since the formation of instructional systems development. In addition, people have experience using ISD in different environments. Lessons from such experiences have been documented. The ISD model can be examined in light of the research and experience that has been collected over the past decade.

It seems appropriate at this time to offer suggestions for changes in instructional systems development procedures to update the ISD model. These changes are more a natural evolution occurring over time than a replacement of the ISD model with a different model. They are offered in this spirit. The recommendations that follow are future oriented, attempting to update the ISD model and to project directions for ISD for another decade.

The recommendations, or predictions, are grouped under four major categories:
1. Future directions of ISD models.
2. Broadening the base of ISD models.
3. Improvements in the ISD models.
4. Likely changes in specific ISD procedures.

The remainder of this chapter is organized into these four areas.

Future Directions of ISD Models

There are several changes that will likely occur in ISD models to bring them more in line with developments that affect education and training. These directions include:

- **cognitively oriented models to replace behaviorally oriented models**

Research on human learning in recent years has been conducted in the framework of cognitive psychology. The emphasis has been on understanding the internal mechanisms that underly human learning rather than just looking at the resulting behaviors. As a result of this research we know much more about the variables that influence human learning. Current ISD models fail to take this new information into account and thus limit the effectiveness of the models. New models will have more of a cognitive orientation in procedures related to task analysis, assessment of entry behaviors, planning of instructional strategies, and design of the lessons.

- **more attention to implementation/acceptance/the user of the model**

Future Directions for Instructional Systems Development

One of the main problem areas with the use of ISD models has been with the implementation of the model and not with the individual steps or procedures per se. Newer ISD models will pay more attention to how they are to be implemented. Concerns with user acceptance will influence how the models themselves are conceptualized.

- more emphasis on how to do it not just what to do

Earlier ISD models represented the steps or procedures necessary for developing instructional systems. Often these models offered no guidance regarding how to perform the steps. Future ISD models will incorporate information about how to do each step, perhaps in the form of examples, job aids, or embedded training.

- improvements in instructional strategy areas

The instructional strategy area has been termed the "black hole" of ISD, since most models had little to say about this area. Current ISD models are very strong on front-end analysis and evaluation but noticeably weak on the mid-portion, on the instructional strategy aspects. There is little guidance on the actual planning and development of the instruction—on the "how to teach" part—in ISD models. Future ISD models will incorporate recent developments in instructional psychology that focus on the requirements for successful instruction.

- separation of the delivery and management of ISD from development

For the most part ISD models have confounded the development of instructional systems with the delivery of the instruction and the management of both the development process and the delivery process. Attempts at incorporating all of this into a single model has caused confusion among users and has resulted in inadequate attention being paid to certain aspects of ISD. This failure to separate management tasks from development tasks has resulted in mismanagement of ISD efforts, which has restricted the potential impact of instructional systems development. By separating these functions, the ISD process can be better managed and the delivery of the education or training can be enhanced. Thus, we should begin to see companion models that describe how the ISD process should be managed and how instruction should be delivered.

- more algorithms on how-to-do-it

As knowledge about the procedures in ISD increases, we should be better able to proceduralize that knowledge. By representing this knowledge in the form of algorithms, users of ISD will be better able to improve their actions. It will become easier to "understand" ISD by reference to these algorithms. There will be many separate algorithms developed for different procedures in the ISD model.

Broadening of the Base of ISD Models

The initial underpinnings of instructional systems development efforts were fairly restrictive. The major influences were educational psychology (especially the behavioral variety including programmed instruction), general systems theory, some work in audio-visual instruction, and a bit from communications theory. As instructional systems development matures, influences will be felt from other areas. ISD models will broaden by incorporating developments from other fields. This influence will result in more elaborate procedures in ISD models. The strongest influences on ISD models will come from anthropology, cognitive sciences, sociology, and organizational behavior. Some specific changes that are anticipated include:

- front-end analysis shaped by anthropological methods

For years anthropologists have explored ways of coming to know a different culture. Some of their methods will begin to be used in ISD to help understand the circumstances or "culture" in

Future Directions for Instructional Systems Development 211

which the education or training is to occur. The greatest impact of anthropological methods on ISD models will be in the area of front-end analysis, particularly needs assessment, constraint analysis, and analysis of the target population. Qualitative approaches to identifying needs will be used along with the more common quantitative approaches now employed. Ethnography will become a tool to help educators and trainers understand the system in which they are working and the constraints that will operate on any instructional system. The use of ethnography is a natural "fit" when working in developing nations; ethnography will prove to be equally well suited to working in large organizations or agencies.

- **design and delivery shaped by the psychology of perception and cognitive sciences**

Recent developments in understanding how humans perceive their environment and how they process information will have an impact on how training programs and materials are designed. Great strides are being made in identifying the factors that affect what we perceive. This research information will be incorporated into the design of instructional displays. Research on information processing in memory will continue to guide how lessons are designed and structured. Work in the area of knowledge representation will serve to guide the analysis and representation of subject matter knowledge. The analysis of differences between experts and novices in a particular domain of knowledge will help define the selection and organization of instructional content.

- **implementation will be shaped by market research**

A key problem that occurs when implementing, or attempting to implement, an instructional systems development project has more to do with acceptance of change itself rather than with any ISD procedure. The idea of building a better mousetrap and seeing the world beat a path to your door is not accurate. In addition to care in carrying out the ISD process, we will see the same care and attention being paid to implementing the "new" instructional system. Techniques associated with market research will be used to gauge and to influence the acceptance of ISD within an organization or agency.

- evaluation will be influenced by anthropology and sociology

The standard methods of assessing learners' mastery of specific objectives will be supplemented with broader assessments examining the impact of the education and training on the organization and the contribution of the education or training to the organization's mission. Broader outcomes will be assessed, by qualitative as well as quantitative methods.

- the design of training programs will be shaped by job design

Considerable effort is underway to engineer improved working environments for employees by examining what they do and how they do it. Changes in the nature of how people work will have implications for how training should be designed and delivered.

Improvements in Models

Several improvements can be expected in the design of ISD models themselves. Certain aspects of ISD models are likely to be revised to reflect changes in what we know about the processes of designing and delivering education and training. The following changes are expected.

- distinctions between macro and micro models

Most ISD models confuse procedures for the development of programs and curricula (macro-level) with procedures for the development of individual lessons and materials (micro-level). Newer ISD models will delineate this distinction and reduce confusion about what is being done at each step in the model.

Future Directions for Instructional Systems Development 213

- **development of alterable ISD models**

It is not realistic to expect one standard model for instructional systems development to be equally useful in all situations. There are too many different variables and constraints associated with different sites for one ISD model to "fit" each case. We will begin to see the development of ISD models that are designed to be altered by the user the same way an adjustable wrench is altered by its user. Adjustments are made to the tool (the wrench) by the user at the time of use so that it fits the circumstances of use (the size of bolt to be tightened). Likewise, ISD models will have some built-in features that allow them to be modified to fit the circumstances of their use. Alterable ISD models should allow for more precision in use and greater use.

- **product driven models**

To date most ISD models reflect the processes to be followed when developing instructional systems. They prescribe the processes to be accomplished, not the products. Some emerging ISD models will put more emphasis on the tangible products. These product oriented models will free the developers to use whatever processes they choose to create the product; they will be more flexible in use. Such product-driven ISD models will require higher-skilled designers and developers because they will have to make more decisions about how to proceed in an ISD project. These models will require professional ISD specialists who can select from among a variety of procedures and apply them as necessary to create an acceptable ISD project.

- **iterative rather than linear models**

ISD models for the future will not be linear as current models are. Rather, future ISD models will be iterative, requiring a certain amount of forwards and backwards movement. Certain early steps will have to be revisited as one proceeds through the model because later decisions or data will affect what occurred at earlier stages.

- **Emphasis on improving on-going training**

Virtually all ISD models are designed to be used in situations where new programs are being developed. These ISD models start fresh, looking at job analysis and needs assessments and proceed through development of instructional materials. These models fail to take into account the reality of their use in situations where training is on-going. Most ISD projects do not begin from scratch but rather attempt to improve existing programs. ISD models will begin to reflect this reality by examining and validating existing products such as task lists, objectives, instructional materials, and tests in an ISD framework to prevent unexamined use of existing course products. Often ISD principles are violated when existing course products are "inserted" during an ISD project.

- **More attention to quality control**

To be effective ISD projects must be managed to ensure that the ISD procedures are followed and that each procedure is executed faithfully. In short, someone must take control for examining the quality of the process and products in an ISD project. ISD models will begin to reflect this need for quality control, especially for large scale projects. Guidelines for assessing quality will accompany ISD models.

- **Models within models that elaborate on ISD procedures**

Future ISD models will be "layered" to show the main procedures at first glance, then to show subprocedures in more detail. Users of ISD models will be able to zoom in and out within a model to see where they are, what happened before and what happens next within the model. They will be able to move from the "big picture" to more detail and guidance on specific procedures.

- **More attention to lesson design**

Future Directions for Instructional Systems Development 215

Most ISD models are deficient with regard to the design of individual lessons and instructional displays. Emerging ISD models will include guidance on aspects of lessons design. Lesson design will become more science and less art leading to more predictable results.

- Graduated models from ideal to quick and dirty

Current models are designed to prescribe the ideal set of procedures for developing instructional systems. Having an idealized set of procedures is essential as a target or norm to which we can aspire. However, constraints often force abandonment of the ideal. Often there is not sufficient time nor adequate funding to complete each step in the ISD process, so short-cuts are taken of necessity. Whole procedures are omitted because of pressures. This compromises the ISD process and likely results in a failure to reap the benefits normally associated with ISD projects. The alternative is an ISD model that incorporates trade-offs at each step. For each step in the ISD model, the ideal is described, but then alternatives are offered if constraints prevent the recommended step from being followed. Of course, the benefits and costs of each alternative are also presented so the developer or manager can see exactly what he or she is trading away. ISD will be presented in such a fashion that it is not an all-or-none proposition. Rather, there will be several graduated procedures for accomplishing each step with the benefits and limitations of each clearly spelled out in advance.

- **More training of managers of ISD projects**

We will begin to pay more attention to the training of persons who will manage ISD projects. Just indicating the procedures to be followed in developing instructional systems has not been sufficient. Problems occur when ISD projects are not managed adequately. Future ISD models will address the management aspects more than prior models. These models will indicate what managers should look for when assessing the outputs from each step. Models will specify not only what should be done at each step but also the criteria for judging the products, an indication of relative time required, quality control points, and management decisions that are required.

- **procedural flowcharts incorporated within models**

In line with more attention on how to do it, ISD models will include procedural flowcharts that depict how to carry out each procedure in the model. These procedural flowcharts will aid the understanding of the ISD model and will aid in the use of the model.

Likely Changes in Specific ISD Procedures

ANALYSIS

- **cognitive task analysis**

In analyzing tasks more attention will be paid to the internal aspects of the task—what is occurring mentally when a job incumbent is completing the task. The first cut at analysis will likely be to identify the behavioral sequences, or procedures, that are followed in completing a task. Then these procedures will be analyzed to determine the information processing requirements for completing the procedures. By analyzing such cognitive requirements, we will be able to specify what must be "known" in order to perform the job.

- **knowledge engineering methods**

In addition to cognitive task analysis, knowledge engineering methods will be used to extract and represent knowledge that skilled incumbents possess. Much of the work on knowledge representation will result in new ways to identify and structure the essential knowledge a new employee must have for him or her to be successful. Work on knowledge engineering will also influence the development of expert systems which will be integrated into the job environment and will impact what an employee must know.

- expert-novice distinctions

Considerable research is being conducted on differences between experts and novices in a variety of content domains. An expert's knowledge is different from a novice's not only in terms of the amount of information possessed but, more importantly, in how the information is organized and used. More is being learned about the transition from novice to expert and what is required for this transition. It is likely that instruction should be different for people located at different points along the expert-novice continuum. Thus the analysis of the knowledge of experts and novices can contribute to the development of appropriate content for training programs.

- pretesting of acceptance

Many problems associated with the use of ISD models have little to do with the models and much to do with the acceptance of the models and the changes they represent from business as usual. Future ISD models will build some form of market research into the models themselves to pretest the acceptance of the model. Several versions may be pretested to determine probable levels of acceptance. Focus groups might participate in model construction to help ensure acceptance.

- expert systems as job aids

The analysis portion of ISD models requires much time and effort from highly trained personnel. The use of expert systems as elaborate job aids will reduce the labor costs for front-end analysis. By capturing the knowledge of expert analysts and representing this in some knowledge-based system, less well trained analysts will be able to complete front-end analyses. Highly trained analysts should be able to work "smarter" using expert systems and be more efficient in completing their analyses.

- data bases of skills

It is likely that we will begin to see more data bases of job skills that can be used when completing front-end analyses. This would allow for more "cutting and pasting" and less reinvention of the wheel. Once the major components, or duties, associated with a job had been identified, data bases could be scanned to see if the subcomponents, or tasks, have been identified for those duties. This would reduce duplication of effort.

DESIGN

- better models for sequencing instruction

The sequencing of instructional content has continued largely in the tradition of programmed instruction, a simple to complex sequence. Recent work in cognitive psychology would indicate that other sequences are more effective in facilitating learning. These models for sequencing instruction suggest starting with the main idea or overall structure—the big picture—and then progressively filling in the specifics.

- new methods for instructional analysis

As more is known about how subject matter content is structured and learned, improved methods for completing instructional analysis will become available. Different instructional analysis methods will be used to analyze different kinds of instructional content. Information will be analyzed differently than motor skills, etc.

- use of knowledge representation methods

Knowledge representation methods being pioneered in the field of artificial intelligence will influence how instructional design is done. This will alter the way lessons are designed and structured. More attention will be paid to the relationships among content to be learned.

Future Directions for Instructional Systems Development

- algorithms for identifying instructional requirements

The research information regarding how to teach different content will be codified in such fashion as to allow development of algorithms to guide designers in specifying what must be done for successful instruction. Thus, planning the strategy and tactics for instruction will become less of an art and more of a science.

- less emphasis on narrowly defined behavioral objectives

Newer ISD models will be less "strict" about the form of the instructional objectives. More attention will be placed on objectives stating content and its organization rather than objectives stating the behaviors that result from acquisition of such content. Less time will be spent developing objectives and more time will be spent planning instructional strategies and tactics.

- more attention to the design of instructional displays

ISD models will attend to the microdesign aspects including the design of instructional displays. The advice in most ISD models when you get to the point of creating lessons is "do it now." Much guidance is given for analysis and evaluation in ISD now; in the future we will see equal guidance on design of displays.

- better methods for matching instruction to learners

ISD models have paid little attention to learners except for the area of prior knowledge. As research advances our understanding of individual-difference variables in instruction, ISD models will incorporate improved methods for matching the instruction to various characteristics of learners.

- use of expert systems as job aids

Designers make many decisions in planning instruction. Future ISD models will include expert systems to assist designers in carrying out the prescribed steps.

- design decisions based on learning research

More of the decisions made during the design phase will be based on research on human learning and fewer decisions will be based on guess work. ISD will become less of an art form that can be practiced by only a few highly skilled people.

DEVELOPMENT

There will be several changes in how instruction is developed. These include:

- more use of job aids and less trained personnel

The development of instruction is too labor-intensive now and requires highly skilled labor. One way to reduce the costs is to rely more on job aids including expert systems to improve the output of less trained persons. As knowledge becomes more stable through research and development efforts, tasks that formerly required skilled employees will be performed by persons with less skill using some job enhancing devices.

- use of "cut and paste" routines or modules

There will be less reinventing of the wheel and more use of off-the-shelf materials in the form of easily modifiable modules. Rather than selecting whole courses or lessons, ISD models will direct the developers to select and modify for their use smaller stand alone chunks of instructional material. Transitional material and some new material will still be required but the amount of material to be developed will be reduced.

- more use of templates for lessons

ISD models will direct developers to use certain templates or models when developing a lesson. For example, Gagne's instructional events model and Merrill's component display theory prescribe the "parts" of a lesson.

Future Directions for Instructional Systems Development 221

- adaptive instructional models

ISD models will incorporate adaptive models so that the instruction that is presented to an individual learner is adjusted in real time as it is delivered. Thus, within a group of learners no two people may complete the same instructional sequence or see the same instructional displays because these will be adjusted to the individual differences among learners. Prior knowledge and current level of understanding are two variables that will be included in adaptive models. There are many other possibilities including anxiety, learning style, personality type, perceptual abilities, etc.

IMPLEMENTATION

ISD models will begin to pay much more attention to the implementation aspects. Several changes are expected including:

- improved interactive media

More of the education and training will be delivered by interactive rather than one-way media. CBT and interactive video are but the beginning of this trend. The hardware to support interactions will be enhanced with better man-machine interfaces including synthesized speech and voice recognition.

- computer-based control of delivery

While much of the instruction will not be delivered by computer, most of the delivery will be controlled, or managed, by computer. The management of instructional delivery can be handled quite adequately by computer, but it is a burdensome task for humans.

- delivery to groups

It seems that ISD always meant individualized instruction in the past. Future ISD efforts will include much more group delivery and interactions among members of the group during instruction. This will be most evident when the training is for jobs that require working in groups or teams. CBT and interactive video will be designed for delivery to groups.

- **higher resolution and lower cost displays**

The quality of the instruction delivered will be enhanced by the development of higher resolution displays. Costs for delivery will reduce as displays become less expensive.

- **less linear instruction and more hypertext**

ISD models will not force instruction to be so linear as is now the case. We will see instruction that is designed to support the learner "jumping around" through a lesson. There will be more learner controlled instruction and more adaptive instruction. Subsequent instructional displays will be sensitive to needs and responses of learners rather than being programmed to advance linearly.

- **integration of instructional delivery with ongoing work**

The distinction of training and work will begin to disappear as training is integrated with work tasks. ISD models will be designed so that training can be integrated with work. Instead of taking time away from the job to attend an all-day session on some aspect of his job, an employee would "jump into" a brief highly tailored training segment during his work. This may be little more than an enhanced job aid or may be a full blown tutorial lesson.

- **delivery of training as needed at the work site**

ISD models will be designed to include the development of training to be delivered on the job site rather than in a formal school. There will be many more options for delivering instructional in a more flexible fashion.

Future Directions for Instructional Systems Development 223

- more concern with user acceptance

ISD models will incorporate market research techniques to help gauge user acceptance. Concern with acceptance will guide the development and application of models. There will be field tests of the models themselves and modifications based on acceptance. ISD models will become more "user friendly."

EVALUATION

The evaluation aspects of ISD models will change in several directions; these include:

- more efficient models of testing

More efficient models and procedures for testing will be included in ISD models to get performance and knowledge results from persons receiving education and training. Thus the amount of effort and time spent on testing will decrease. Tailored tests are a step in this direction.

- better use of sampling techniques in formative evaluation

Rather than testing large numbers of persons from the target population, formative evaluation procedures will rely on controlled samples to estimate the effectiveness of instructional materials. This will be akin to sampling in opinion polls. Accurate population estimates can be made with small, but carefully constructed samples.

- routine tryout for achievement and acceptance

The development and use of improved evaluation procedures will enable ISD models to include frequent assessments of instruction as it is being developed. This has always been a desire in ISD projects but because of the time and effort evaluation has required, evaluation was often "skipped over." Evaluation will become more of a routine part of ISD.

- automated test item development

Developments in test construction will include computer generation of test items according to specifications. This will be influenced by developments in artificial intelligence and will require some form of machine intelligence.

- computer based test administration/analysis/interpretation/reporting

The tasks associated with the administration, scoring, analysis, interpretation, and reporting of tests will be computer based. Much progress has been made on these tasks and results of these efforts will be incorporated into ISD models to ease the evaluation burden of earlier ISD models.

Summary

No prediction of the future is totally accurate. The very nature of predictions is "guessing at" unknowns. Many assumptions are made which might not hold at a later point in time. So it is with the prediction of the future of instructional systems development. These predicted directions are fairly short-term. Many are unfolding now; the remainder are likely to emerge in the next few years. None of these predictions seem outlandish nor do they require major breakthroughs in technology. These seem to be a realistic set of predictions about where ISD is headed, about where ISD should head. Our hope is that many of these predictions will prove to be true in the next 5 to 10 years. Education and training will be all the richer for it.

Appendix A

Large Scale ISD Programs: Two Case Studies in the Military Services

Robert K. Branson[1]
Florida State University

Introduction

The case studies presented in this chapter are intended to illustrate the application of ISD concepts and procedures in the Army and Navy. Two large scale programs will be discussed to provide a means of contrasting alternative ways to achieve similar goals.[2] The Navy example is the Job Oriented Basic Skills (JOBS) program (Harding, Mogford, Melching, and Showel, 1981), and the follow-up evaluation studies done by Baker and Huff (1981) and Baker and Hamovitch (1983). The Army example is the Job Skills Education Program (JSEP) described by Anderson (1985), Branson and Farr (1984), and Farr (1986).

These two programs are intended to prepare new accessions to the Army and Navy to perform adequately in their initial technical training courses and on the job. The all-volunteer force concept requires all services to compete with civilian employers for members of the available personnel pool of 17-20 year-olds. When the civilian economy is good, the armed forces have more trouble in recruiting than when the economy is soft, but the military mission does not change correspondingly based on changes in the economy. The services must counter all threats and do the same highly technical operations, maintenance, and repair jobs regardless of how well qualified the recruits are. Since these less capable recruits often do poorly in the technical schools, the services have initiated research and development programs conducted by their in-house laboratories to provide compensatory instruction in those identified deficiencies in prerequisite competencies and basic skills.

Probably the most widely cited functional basic skills effort was the Army's FLIT (*F*unctional *Lit*eracy) program developed by Sticht, (1975) when he was with HumRRO (see also Duffy, 1985). The success of FLIT in preparing soldiers to read the materials they were required to use on the job was responsible, in part, for stimulating additional functional basic skills R & D activities by the U.S. Army Research Institute (ARI), the U.S. Navy Personnel Research and Development Center (NPRDC), and the U.S. Air Force Human Resources Laboratory (AFHRL). The results of this current work have only recently begun to be reported at professional meetings (e.g., session 11.28 of the 1986 Annual Meeting of the American Educational Research Association in San Francisco, chaired by Frederiksen; also, session 41.07 chaired by Wilson).

The JOBS and JSEP programs were selected for this chapter because they emphasize different aspects of ISD models to accomplish the intended outcomes. The JOBS instruction is provided in a 30 hour per week traditional lockstep schedule, while the JSEP instruction is accomplished on two computer-based instructional (CBI) systems, MicroTICCIT and PLATO, and with necessary paper and media-based support. The point to be made in this comparison is that in the ISD approach, analysis, design, and management are critical to success—instructional methods rarely are.

An outcome gap had been identified in both the Army and Navy between available manpower capabilities and required capabilities. Given a choice, the services would have first used *selection* as a means of closing the gap. It is far easier and less expensive to select qualified recruits than it is to provide remedial or compensatory instruction. While great progress has been made in improving selection techniques during the past 20 years, the fundamental requirement for a successful selection program is that there must be far more applicants than there are positions so that the less qualified can be turned away. Since selecting better qualified recruits was not an adequate solution, training was chosen as the optimum tradeoff.

Large Scale ISD Programs: Two Case Studies 227

In the next sections of the chapter, details on the specific characteristics of JOBS and JSEP will be discussed in the context of the ISD model, since both are instances of ISD process. Features of each program will be used to highlight different phases of the ISD model.

Analysis
Function of the military personnel system. All potential recruits to the military services are classified for occupational choices based on their scores on the Armed Services Vocational Aptitude Battery (ASVAB). This test provides scores that estimate the probability of success in any of the technical schools to which they could be assigned. Each school has established a profile for selection that yields a success/failure ratio designed to provide a suitable number of graduates to the operational units. The military personnel system must allocate these school assignments to provide an optimum mix of graduates for each job. There is a finite number of billets in each service. Each must be filled with a person who is at least marginally competent to do the job. In that zero-sum environment, if more of the competent people are assigned to one job, fewer can be assigned to others. All of the top quality people cannot be bunched in one occupation, they must be balanced across the force to maximize operational readiness. Maximum readiness is caused by optimum selection, training, and management.

Within each service, the personnel assignment command monitors scores and success in all the schools and makes decisions about the relative quality of recruits and the demands of the schools. In addition, the internal laboratories (AFHRL, ARI, NPRDC) conduct research projects to find alternative assignment and instruction programs to meet the needs of all of the occupational groups. Because of the large number of people who enter the service, the scores and results are constantly available for analysis. When it appears that some kind of selection or training intervention would improve the performance of the recruits, these laboratories go to work to find a solution.

Needs analysis. Identifying performance problems in the operational units of the Army is a routine, ongoing process. Each service has regular large scale operational tests of combat readiness. When important problems are found as a result of those tests, it is often the service laboratory that must design a solution. Since these operational exercises are extremely complex, and data on unit effectiveness is constantly being generated, there is always a systematic process in place to identify the problems. Performance indicators are designed into the system so that flags are raised when results are not adequate to meet operational needs.

When the JOBS program was considered, NPRDC made a careful analysis of service school results, job performance, and retention of service members in the fleet. When they found that the number of recruits in the higher mental categories was not sufficient to meet the needs of the fleet, they proposed a program of compensatory instruction that would help overcome the knowledge and skill deficiencies that prevented the recruits from completing technical schools successfully. Recruits are generally assigned to one of four mental categories, I, II, III, and IV, corresponding roughly to the four statistical quartiles, based on ASVAB scores. A Category I recruit would be expected to succeed in any of the schools or jobs in that service, while a Category IV would be expected to succeed only in a limited number of schools and carefully supervised jobs.

Since compensatory and remedial programs in the armed forces are always designed to serve specific audiences, profiles are developed to establish the zone of eligibility as defined on one or more chosen variables. Compensatory programs are intended to enable the lowest quartile (Category IV) recruits to achieve minimum acceptable performance. The mean scores of the recruits who qualified to attend Navy technical schools was about 28 points higher than those eligible for the JOBS program. Some 4500 candidates were found to be eligible for the program. JOBS eligibility was based on the entrance requirements of the technical training school they were being considered for. If candidates were below the cutoff scores for waivers to these schools and were in the range of eligibility for consideration for JOBS, they were invited to attend JOBS training. About 50% of those eligible volunteered. Their willingness to join may have indicated that they recognized their deficiencies in prerequisite competencies

Large Scale ISD Programs: Two Case Studies 229

and basic skills and may have suggested greater motivation to succeed. Presumably, those who did not volunteer simply could not face any more 30 hour weeks in the schoolhouse.

Task analysis. All military services have models and procedures for identifying the tasks performed on the job (see Branson, Rayner, Cox, Furman, King, and Hannum, 1975). When these tasks have been identified through the selected *job analysis* method, they are then broken down into their subordinate knowledge and skill components through task analysis. In the JSEP project, a contractor (RCA Education Services) was hired by the Army's Training and Doctrine Command to perform the job and task analysis. To reach 85% of the soldiers, 94 jobs had to be analyzed. The ultimate purpose of the task analysis was to find the subordinate basic skill components and prerequisite competencies required to perform those tasks in Army jobs at the lower apprentice skill levels. New soldiers perform tasks requiring lower levels of skill until they can be trained to do more complicated tasks.

The task analysis set out to find instances of basic skills used in each of the selected tasks. When these basic skills were identified, they were classified into a taxonomy containing some 200 statements, each identifying a prerequisite competency. A prerequisite competency was defined as a knowledge or skill subordinate to task performance. Because the tasks of one job are often vastly different from another (e.g., radar repairer vs. tank gunner), soldiers would be given instruction only in those prerequisite competencies that were required on their jobs. Table 1 lists prerequisite competencies from the taxonomy developed by RCA.

Contemporary thought in the field of job and task analysis is to be highly specific about requirements. Rather than ask instructors whether a given job required the use of "mathematics," during the task analysis, the analysis found specific instances of exactly what kind of mathematics was required by collecting *indicator statements* which provided examples of how the prerequisite competency was used in that specific job. To provide some notion of the project's magnitude, the computer printouts of the RCA analysis required some 50,000 pages. The instructional designers were expected to master the use of these materials as source documents to obtain a rich array of examples and con-

Table 1 Selected Prerequisite Competencies from the Taxonomy Developed by RCA.

Numbering and Counting

1a. Match numerals with word names and models

1d. Identify a number which is greater or lesser from a set of numbers

Linear, Weight, and Volume Measures

2a. Interpret the markings on a linear scale

2d. Identify measures of weight (ounces, pounds, grams), pressure (pounds per square inch), and torque (foot pounds)

Degree Measures

3a. Identify degrees and mils as units in determining angular measurement or temperature

Time-Telling Measures

4d. Determine equivalent dates from one calendar form to another using Gregorian and Julian calendars

4f. Convert to Zulu (Greenwich Mean Time)

Fractions/Decimals

14a. Estimate fractional lengths, distance, area, and volume

Graphing in the Coordinate Plane

17b. Specify the digit coordinates of any intersection of lines on a military map

Reference Skills

27d. Locate the title, page, paragraph, figure, or chart needed to answer a question or to solve a problem

Tables/Charts

28c. Use a complex table or chart requiring cross-referencing within or in combination with text material outside the chart

Large Scale ISD Programs: Two Case Studies 231

Table 2 Prerequisite Competencies and Illustrative Indicator Statements from the RCA Job and Task Analysis.

P.C. 4lb **Use and Interpret Hand and Arm Signals**

- Signal operator to swing load out when clear of vehicle
- Clasp hands together to have driver stop tank when lifter is straight up and down over connector with road wheel off the center guides

P.C. 15a **Draw Plane Geometric Figures**

- Draw the area of operation on the map using the platoon leader's map as a reference
- Draw a triangle over grid intersection

P.C. 02c **Measure Lengths of Objects or Distances Using a Ruler, Yardstick, Meter Stick or Scale**

- Determine the centerpoint of the FPF
- Measure along azimuth, the estimated distance
- Read micrometer

P.C. 19d **Use Tables of Trigonometric Functions (Degrees)**

- Extract log sin for value of ½ (Latitude of station-apparent declination of sun)
- Determine log sin vertical angle

texts of use for the lessons. More will be said about that process later. Table 2 provides examples of indicator statements associated with selected prerequisite competencies.

While a different job analysis process was used in the JOBS program (Harding, Mogford, Melching, and Showel, 1981), the information developed served the same functional purpose as the RCA analysis did in the JSEP program. Rather than establish a computer-based management system, the Navy proceeded with the assumption that "job-oriented" basic skills would be specific to a particular type of occupational specialty. Given that the development of a separate JOBS program for each Navy rating would be cost-prohibitive, the Navy attempted to identify clusters for strands of ratings likely to have the same or related basic or prerequisite skills. Of the clusters identified by Baker (1978), the following four, representing 13 ratings (occupational specialties) were selected for inclusion in the JOBS pilot test: Propulsion Engineering, Administrative-Clerical, Electronics, and Operations.

The purpose of the JOBS program is to enable non-school eligible personnel to enter and successfully complete a Navy technical school. As such, the "job" to be analyzed, from which training objectives would be derived, was not the task to be performed at a work station, but, rather, the follow-on technical school. Hence, the skills and knowledges required by students prior to entering any of the technical schools served by a cluster or strand were identified by examining each school's training objectives and materials and by interviewing both faculty and students. As detailed by Harding, Mogford, Melching and Showel (1981), the focus of both the interviews and the examination of course materials was on identifying skills and knowledges that were needed but not taught in the technical schools, or that were taught but were difficult to learn.

The skills and knowledges identified were incorporated into a test battery administered to both JOBS eligible and technical school eligible candidates. The results of this testing provided the basis for the selection of most of the skills and knowledges included in the JOBS curriculum. A skill or knowledge was included if (1) it was not taught in the technical school and 50 percent or less of the JOBS eligible candidates could answer the corresponding test item correctly, or (2) it was taught in the technical school, but 50 percent or less of the school eligible candidates could answer the test item correctly. In the latter case, the difference between the percentage of technical school eligible and JOBS eligible candidates answering the test item correctly had to be at least 15 percent.

While a computer-based management system might have provided what would appear to be a theoretically better assignment system, evidence from the JSEP program suggests that, as a practical matter, a limited number of grouped curricula may serve just as well.

Design

The purpose of the JOBS training program was to improve the chances that recruits would be successful in their Navy jobs. Success can be defined in numerous ways, depending on whether more proximal or distal criteria are used. The first variable of interest is whether the recruit can meet the standards of the basic skills instruction. Next, are the trainees who had the JOBS instruction just as likely as those who are school qualified to graduate from the service school? Third, how likely is it that those who have had the instruction are retained in the Navy as long or longer than those who are school qualified? Fourth, does the instruction seem to have any effect on discipline and job performance? Baker and Huff (1981) provide a complete listing of the variables of interest used in the JOBS evaluation.

Having chosen training as the optimum method to solve the problem, it was then necessary to design a training system that would provide effective and efficient instruction. There are two fundamental and important differences between military training and public education: (1) It is not the military mission to provide basic skills instruction for its own value, rather, it is provided only as a means of achieving job performance, and, (2) It is necessary to provide the same training to all members of the service who hold the same job.

External influences. The General Accounting Office (GAO), the audit agency reporting directly to Congress, regularly audits military training programs to see if they are meeting their stated expectations. One basis on which the GAO evaluates training programs is on the degree of standardization from one military installation to another. The GAO takes the position that if basic skills programs are required to solve problems, there ought to be documentation to show the relevance of the training to the mission of the service. Basic skills improvements, for their own sake, are not considered mission oriented.

Recent GAO reports have identified certain basic skills programs in the Army that varied greatly from one installation to another around the country and the world (U.S. General Accounting Office, 1983). In response to that report, the JSEP design goals included the requirement for a standardized curriculum that could be uniformly delivered regardless of the location. Such uniformity virtually cannot be achieved with standard classroom instruction unless a great deal of control is exercised over the instruction. That level of supervision is almost non-existent in education-related jobs, military or civilian.

The design goals of JSEP were established by the analysis of problems discovered in the field, in reaction to external audit reports, and in the context of available and emerging technologies. One reason that CBI was chosen is that it provides a means to deliver identical instruction at numerous remote sites from the same data base, and standardization of testing and completion criteria can be identical. This design feature is critical in many organizations, in particular where participation in the training may have some future impact on retention in the service or promotion.

Costs. One of the more controversial issues today is that of whether mediated or computer-based instruction is cost-effective. No instructional approach is, in itself, more or less cost-effective. While alternative designs may be *a priori* more advantageous, the principal determinant of cost-effectiveness is management's commitment to achieve cost goals. Ineffective managers and powerful political influences can make the most efficiently designed instruction hopelessly expensive, and good managers can make the most primitive forms of instruction pay off handsomely. Technology contributes to cost-effectiveness only when it can be fully exploited in the social and political climate of the organiza-

Large Scale ISD Programs: Two Case Studies 235

tion. There is no more revealing story of the mind-set of conservative and traditional educational administrators than that told of having ". . . all of the technology bought by the government locked up in the closet."

Standardization. The design feature of the JSEP and JOBS programs that appears to have met all of the external criticism is that of standardization of curriculum. When a careful and deliberate analysis of the problem to be addressed has occurred and instruction has been designed to overcome identified deficiencies, individual instructors must be prevented from reworking the curriculum to fit their own personal, and often naive, beliefs about ". . . what the soldiers need." The whimsical curriculum decisions made by poorly supervised teachers in the public education system is probably one of the significant contributors to the crisis in education.

The typical journeyman instructor, whether civilian or military, is bound by limited experience, which causes them to believe that some aspects of the curriculum require more emphasis than the documented needs analysis establishes. It is simply not true that individual teachers, no matter how sincere and well motivated, always "know what's best" for the students.

Management. If the recommended implementation and management plan for JSEP is carefully followed and the instructor-to-student ratio is responsibly managed, then the program will result in important improvements in the quality of instruction, and the total cost of ownership will be less over the life cycle than methods currently in use. If, on the other hand, the individual managers choose to modify and change the program in such a way that more instructors are required, or do not choose to exploit fully the efficient design features, the total cost of ownership will be greater, even though JSEP performance results will probably be superior to conventional instruction.

If anything is known about the impact of technology on the costs and effectiveness of instruction it is that management of the implementation is the critical element. Technology reduces costs only when it replaces alternatives. During the economic depression of the late 1920's and 1930's, a labor representative and an engineer were watching a steam shovel chew huge bites of earth from an excavation and empty them into a large dump truck. The labor representative said that it would be better for the workers if the steam shovel could be replaced with a hundred men with shovels. Recognizing that using a hundred men was not economically feasible, the project engineer responded that it might be even more wonderful if it were replaced by a thousand men with teaspoons. No matter how potentially productive the technology, the management of implementation is the critical element in establishing cost-effectiveness.

The curriculum. The JSEP curriculum consists of some three hundred lessons on each of the two computer systems. The lessons are divided into two major groups: Diagnostic Review Lessons and Skill Development Lessons. The Diagnostic Review Lessons serve the purpose of giving a brief review of the material to stimulate recall, then give the lesson test. If soldiers pass the lesson test, they do not have to take the much longer Skill Development Lesson. The basic assumption underlying the design is that the program is remedial—soldiers will have already had the instruction in previous schooling, and a brief review is sufficient to restore it to working memory. If they do not pass, they then take the longer version of the lesson that provides complete instruction.

In addition to the JSEP computer-based curriculum, a significant component of the materials was paper-based. These paper-based lessons are computer managed and integrated with the rest of the lessons. While it was intended that the maximum possible amount of instruction be delivered by computer, there are certain basic skills and prerequisite competencies that cannot be validly taught on line. Sketching a layout for an encampment, following oral instructions, and producing outlines are much better taught off-line.

Development

How managers choose to organize an instructional development project depends on the resources and support available. Two major management structures are generally considered. In the first, subject matter experts (SMEs) are trained to do the development under the supervision of the instructional systems profes-

Large Scale ISD Programs: Two Case Studies

sionals. In the second approach, instructional designers do the majority of the work, depending on the SMEs for content validity. Successful projects have been accomplished using both methods. If SMEs are highly paid or in short supply, it probably makes more sense to have the instructional designers do the work and coordinate it with the SMEs. If the SMEs are available at reasonable cost, it may be more efficient to train them to do the design work. Generally literate SMEs can usually be taught all they need to know about instructional development for a specific project in two or three weeks, provided that they will do the work under the supervision of experienced instructional systems professionals.

JOBS organization. The core of the JOBS organizational structure consisted of four team leaders-instructional designers, one for each strand or cluster, supported by instructional developers-writers possessing varying degrees of ISD experience. These personnel, as well as a project director, a project manager, and subject matter experts in military training were all from Northrop Services, Inc., the development contractor for the JOBS program. Subject matter experts were also provided by the Navy to assist Northrop Services, Inc., in the development effort. The NPRDC served as the Navy's technical representative and with the assistance of approximately 15 apprentice instructional designers secured under contract from the San Diego State University, reviewed each document produced by Northrop Services, Inc.

JSEP organization. The JSEP development approach used apprentice instructional designers to do the actual writing under the supervision of journeymen. This organizational structure was chosen primarily because Army SMEs were not available at the development site. Most of the subject matter expertise was obtained from basic skills specialists at Florida State University, the Army literature provided, and from the job and task analysis documentation obtained from RCA. Consultations with individual SMEs was rarely possible under the terms of the contract. Teams were formed to approach various groups of prerequisite competencies. Representatives on the teams were from the group of instructional designers, curriculum area experts (reading, arithmetic, geography, etc.), and those having Army expertise. This latter group consisted of veterans with recent military experience, professionals with extensive military training experience, and ROTC instructors who were available on the Florida State University campus.

The developers of the JOBS program also indicated that they felt the need for additional time with the SMEs and instructor group (Harding, Mogford, Melching, and Showel, 1981, p. 16). While developing curriculum away from the implementation site may be advantageous, it is clear that improvements can be made if SMEs and users can be more readily accessible.

Content validity. Instructional systems development is a discipline principally concerned with translating the knowledge of the past into forms where it can be efficiently acquired by new learners. As in any translation, complete accuracy is the prime requirement. Accuracy in subject matter areas and technical fields is referred to as *content validity*. There is no more difficult task in developing instructional materials than to insure that they have excellent content validity. To teach well that which is factually or conceptually wrong, no matter how clever the design or well intentioned the designer, is unethical. That achieving content validity is a difficult process never excuses the designer from the requirement.

The design team. The lesson series treatment (the overall plan, approach, and description) was discussed by the design team and reviewed by content specialists who were not military SMEs. When the design approach had been agreed upon, individual lessons were then assigned to designers. It was the role of the supervisor to insure that there was coherence among the individual lessons within each series and to coordinate with other supervisors across lesson groups to insure that there was no unplanned redundancy. The first draft of the lessons was completed by a team consisting of a designer and subject matter advisor, then it was reviewed by the supervisor and by the Army experts to insure that it was consistent with current thought. These Army reviewers (called "greeners" because they painted the content Army green to insure functional relevance) found instances of the use of the prerequisite competency in various military jobs by searching Field Manuals and Technical Manuals in the library and the task analysis printouts containing the indicator statements mentioned earlier.

When the initial draft of the lesson was completed, it was sent to ARI for review and critique. Comments made by the reviewers were put in priority order and incorporated in the lesson specifications. When all comments and corrections had been completed, the lesson specifications were sent to the Micro TICCIT and PLATO supervisors who assigned them to programmers for development. When necessary, these programmers, who were trained in the use of the authoring languages of their respective computer systems (ADAPT for MicroTICCIT; TUTOR for PLATO), consulted with the supervisors on the technicalities of the lessons.

Large and small projects. There are fundamental differences in design strategy between large, complex projects and individual course development efforts. Most authors of ISD models have neither managed nor participated in the development of large scale, systematic curriculum development projects. Andrews and Goodson (1980) reviewed some 40 ISD models published by as many authors who write extensively for the professional literature—university professors or laboratory scientists—most of whom have only limited experience in designing and managing major projects.

ISD models can be roughly classified as either course-based or system-based. Course-based models are used by individual authors who typically concern themselves with developing only one independent course in which they are the SME. System-based models are used on larger projects by teams of SMEs, designers, and other specialists to produce an integrated curriculum.

Individual professors or technicians who learn popular design models or authoring languages to develop their courses have a totally different perspective on how courses should be developed than managers who must coordinate the work of many different professional groups. Configuration management, making sure that all elements of the system are current and consistent with each other, is the most difficult problem for managers of large programs. It is a trivial consideration for individual authors, since they are both authors and SMEs on the same lessons. Consequently, individual authors will have completely different needs and perspectives about the models they choose.

Implementation

In large projects, implementation is a major issue (Branson, 1979). In addition to the conceptual problems of putting a program in place and making it work, there are logistics issues that occupy considerably more time than most developers of individual courses would imagine. Because large organizations are bureaucratic, implementation plans must be developed well in advance and contain explicit details to minimize errors of interpretation. Not because there is any conspiracy of resistance, but because different parts of the work must be assigned to different departments within the receiving organization, such as facilities rehabilitation, maintenance, telecommunications, and security (see Back and McCombs, 1984).

The need for this level of precision can be illustrated by an experience reported by Gagne (1971) to have occurred during the second world war. Army Air Force psychologists were assigned to develop selection tests to screen good pilot candidates from those with more limited abilities. To do that, testing apparatus had to be built from raw materials, since none were commercially available. The chief psychologist placed an order for a lathe to be used to turn raw stock into joysticks resembling those found in aircraft. When the purchasing agent called him back to ask how large the lathe should be, he looked at the mock cockpit and estimated the joystick to be about two feet long. He told the buyer to get him a lathe that would handle about two foot raw stock.

Several months later, the receiving department called him to say that the lathe had arrived, and asked him who was supposed to get it off the railroad car and into the psychology building. Astonished, the lab director dashed to the receiving department only to see a lathe on a flatcar capable of working two foot stock—two feet in *diameter*.

Top down management. Large projects are almost always funded at an organization level considerably higher than that at which implementation is planned. In JSEP, the program manager is located in a central headquarters at the Department of the Army level, while implementation is planned for local installations within the 13 major commands in the continental United States. This is a classic case of centralized design and development with planned implementation at local levels. Most often, one hears at the local level that some new program is ". . . being imposed from above," or, more graphically, ". . . they're trying to shove that program down our throats." The presumption behind these remarks is that the local folks can handle the problem just fine and they don't want any changes caused from above. The facts are that the local folks are not doing just fine or it is unlikely that the program would have been developed in the first place. It is essential for the government's program manager and the contractor's principal investigator together to negotiate a *rapprochement* between these two polar opposite perspectives: central versus local management.

Because local organizations are typically satisfied with their own performance, they will not seek research-based methods to improve. Fads will come and go, but fundamental improvements cannot be made locally due to the political climate that maintains the status quo. Any program designed to produce fundamental change that is developed elsewhere for implementation at local levels will be criticized and sabotaged regardless of the efforts to apply change agent strategy. Innovations and modest changes that can be implemented in the present context of local operations and that do not require changes in the infrastructure can be much more easily implemented with effective change agentry. Individual classroom computers represent an example of modest changes that can be implemented easily. Differentiated staffing and peer tutoring models represent examples of research-based changes that probably cannot be institutionalized in schools due to the antiquated management structure.

Implicit and explicit models of instruction. Regardless of how many new research-based programs for improvement in public education are offered, they are regularly rejected at local levels. If the same kinds of practices were followed in medicine or engineering, there would be a landslide of lawsuits charging malpractice. Keep in mind that all of the people who manage the organizations in which you will implement ISD programs will be most likely to have a strongly traditional concept of instruction, called an *implicit model* of instruction. It is implicit because one learns the proper traditional model incidentally to learning the subject matter.

ISD models are called *explicit models* of instruction because they clearly indicate both the form and content of instruction in the specifications. Systematic models nearly always require approaches to instruction that differ from the traditional school model. Managers and executives are not likely to have experienced ISD courses. Most of them will have gone through traditional curricula in elementary, middle, high school, and college, and they will consider any variation in the implicit model of instruction or management to be abnormal or improper.

Contrasting models. There were major differences between the JSEP and JOBS programs in the implementation process. Remember that JOBS was a lockstep program offered in the same training facilities and required modifications in curriculum content only. While no major changes were required in the management structure, a considerable effort had to be expended to insure that the instructors were adhering to the specified curriculum. In contrast, JSEP is a CBI program requiring extensive logistical support, difficult scheduling of soldiers, and more complex management.

Before the computer systems could be installed at selected posts, the facilities had to be modified. In the military, work requests that require the acquisition of capital equipment and facilities modification require a long time to be completed. The post engineer, who must accomplish all facilities work, does not have the same schedule of priorities that the instructional staff does. Consequently, one is faced with either a delay in implementation or requesting the higher level commanders to change the engineer's priorities. To do the latter takes much courage since such requests always raise the visibility of the program to higher command levels before it has been tried out, revised, and fine-tuned for operations.

Field tests. The JSEP program was field tested at four sites, two with the PLATO system and two with MicroTICCIT. The PLATO system was installed at Fort Leonard Wood, MO, and Fort Sill, OK, while the MicroTICCIT system was installed at Fort Lewis, WA, and Fort Riely, KS. Each of these installations is commanded by a two-star general who has virtually complete line authority and responsibility for management of all activities on the posts. Computer-based instructional systems for education are not a high priority for any of them.

The agency sponsoring development, the Education Division of the Army's Office of the Deputy Chief of Staff for Personnel, is a staff function and may not give direct orders to the tryout sites. Large scale programs always meet with this division of line and staff responsibility. Consequently, communicating the intentions of the program to all levels and commands is essential even to achieve the commitment to conduct tryouts at local posts.

Selecting sites. The four field test sites were chosen by negotiation with their major command headquarters. The tryout program was announced to all the Army Education Centers in the US and sites were chosen or volunteered their services. From the range of those who were considered, the final selection was made. But, why would sites choose to be selected? In the Army, as in civilian organizations, the most plausible reasons are visibility, professional pride, and challenge—to have accepted the responsibility for a large additional assignment was not a necessary part of the job. The Education Services Officers at these posts were not relieved of any of their other duties and responsibilities.

Implementation of JOBS. The four JOBS courses were implemented at the Naval Training Center, San Diego, using civilian contract instructors employed by the San Diego Community College District. To gain detailed information about JOBS implementation, the Navy also employed a number of university students as classroom observers. These observers used course materials as guides and noted deviations from the intended plans. They also recommended improvements to the program.

Training. Because JSEP is intended for implementation in a self-paced learning center environment and soldiers are managed through the curriculum by the computer, it was necessary to provide extensive instructor and manager training before starting soldiers through the program. The training took place about a week before the first soldiers started through the materials. The Florida State University JSEP field coordinator stayed at each tryout site for the first week to be sure that the local staff could operate the program effectively. Substitute instructors were also trained so that there was ample backup in staffing when absences occurred.

Evidence gathered by the classroom observers during the JOBS implementation suggest that there were numerous deviations from the instructor guides. Instructors omitted certain segments of the curriculum, modified others, and inserted activities not specified in the instructors' guides. Since these lessons were carefully designed, based on a thorough job analysis, the results were not as positive as they otherwise might have been.

JSEP is intended to be a continuing program in the Army for as many years as the current requirements exist. Plans were made in the mid 1980s to deliver the instruction on the Army's new Electronic Information Delivery System (EIDS: a microcomputer-based videodisc workstation) as that system is brought on-line in the late 1980s and early 1990s. Plans are being put in place to provide for adding to the contractor-developed curriculum and revising it as changes occur within the Army.

Evaluation

Rarely does one attend meetings of professional societies where the problem of conducting adequate program evaluation in field settings is not discussed. Managers in the military and in business do not see their purpose as collecting data for program evaluation. Rather, they see themselves as being there to *make things happen*. This thrust on the part of managers to develop and implement programs before they have been systematically evaluated has been a source of extreme frustration to instructional psychologists for generations. I believe that it will continue to be so for generations to come. We must develop creative approaches to evaluation that allow for reasonable decision making even without the precision that we would like.

The purpose of all instruction in the military is to enhance careers and improve job performance. The knowledge and skill required to operate, repair, and maintain equipment, and the attitudes and heritage of the profession of arms, accumulated over thousands of years, must be transmitted to the new generations effectively and efficiently. In military training programs, it is necessary to measure both training effectiveness and its relationship to job performance. In basic skills and compensatory programs, it is further necessary to assess whether the preparatory instruction contributed to the improvement of training effectiveness. The chain of evidence must be convincing if the programs are to be implemented and kept in place over time.

In the next section, the evaluation discussion will be centered on the formative issues for JSEP and the summative issues for JOBS.

JSEP formative evaluation issues. Perhaps the most difficult issue to manage is the discrepancy between the customer's notion of formative evaluation and that of the developer. All the design models reviewed by Andrews and Goodson (1980) included a major section on formative evaluation. All of the leading textbooks in the field indicate that the only way to make instruction work is to try it out on the target population and revise it based on the empirical tryout data.

In a career spanning some twenty five years of working on both small and large ISD projects, I have yet to negotiate an opportunity to conduct the kind of formative evaluation conceptualized in the textbooks and models. Even though departments and sites volunteer to "test" materials, their expectations are that the materials will be completely finished before they are brought to the test site. As a result, implementation is often hindered by the lack of data upon which to base revisions.

But, why discuss implementation in the formative evaluation section? Because that is where most large projects conduct the formative evaluation. We all know that ISD models require that small-scale tryouts occur before the instruction is implemented. However, when one is attempting to try out a large computer-based instruction project, it cannot be tried out without first being implemented at the tryout site. The target audience is available only at the site. This dilemma represents a clear distinction between the problems encountered in course-based models and system-based models.

Target audience. A second requirement of the models is that the tryouts be conducted on members of the target audience that have the required entry skills. From a logical and model-building design perspective, that is the way that it should be. The problem is in obtaining these people when you are ready to test them. Only in small-scale course development projects is the target audience available. In large organizations, two or more departments must coordinate their work schedules and priorities to provide the people for the test. It is far more likely that those on whom the materials are tried will come from the available manpower pool rather than the specified audience.

In the Army, nothing is static. Soldiers go through training programs, are assigned to jobs, go on field training exercises, go on standby alert, and on personal and administrative leave. They are not sitting in an auditorium or barracks waiting to be called to the test. In JSEP field trials, the attrition rate of soldiers was more than ten percent per week. No dropouts were caused by poor performance, only by legitimate military requirements or rights. You can expect about the same attrition rate in government and industrial training programs as well.

Some years ago in a discussion about conducting tryouts of instructional materials on a military base a young faculty member, who was striving intensely for tenure and needed carefully drawn experimental and control groups for her study, had a terrible show of temper when I told her about the realities of attrition in these kinds of studies. I said, "I'm sorry, but I cannot assure you that the same subjects will be available for the entire four-week period."

In absolute frustration, she responded emotionally, "You just go in there and *insist* that they remain in the experiment."

"I would be glad to do that," I replied calmly, "but, I haven't the vaguest idea whom to insist *to*."

Hardware and software problems. Computer systems fail. Lightning strikes, air conditioners quit working, disc drives don't, sprinkler systems do, and many other catastrophes can happen. Systems don't fail often, but when they do, it is not good practice to send the soldiers back to their units in the middle of the day. Accordingly, lessons must be on hand to provide alternatives to the computer when the system is down. Sometimes it is down for only a few minutes, sometimes it is down for more than a day.

Collecting system reliability data is a critical part of the formative evaluation. These data provide the inputs to calculate two important operating statistics: (1) Mean time between failures (MTBF), and (2) Mean time to repair (MTTR). Estimates of these variables are used to plan the number of systems required, the number of spares, and are indicators of the amount of non-computer courseware that must be on hand to continue the instruction during the down periods.

Soldier preferences. The relationships between student preferences for learning materials and performance in the courses are tenuous at best and are more often negative than positive. Students do not always prefer those methods that cause them to learn best. Preference data is not collected to see whether soldiers learn better from one method than another, but to find out whether the proposed system is marginally acceptable. It requires intense management attention to get people to do tasks they do not want to do. If there are many complaints from soldiers, the system ultimately will be withdrawn, since CBI is not a part of the Army's true mission. While the Army is quite willing to make people march 25 miles regardless of how much they really want to, it is not willing to put that same effort into enforcing unpopular educational methods. Neither are the public schools, regardless of how much improvement the new methods bring.

In the JSEP program, the soldiers found the CBI systems to their liking (see Peterson and Farr, 1984). What complaints they had about the systems were primarily concerned with glitches in the lessons and errors on the tests. When these were corrected, the complaints were reduced effectively to zero.

Test instruments. All tests administered to soldiers must have command approval before they are used. According to Army regulations based on the Privacy Act, soldiers may refuse to give certain information. The soldiers must be told what the tests are used for and whether they are a part of the permanent record. When they are given a 150 item test for the purpose of improving the quality of the test items and establishing entry skills, they know in advance that their scores on the test are not important to their careers. Trying to administer a difficult four-hour test under those conditions and collect respectable data gives the contractor a true opportunity to excel.

In addition to the summative tests, there were also individual lesson tests signifying whether soldiers successfully passed a lesson. Opinions vary about how many people should go through a lesson at each stage of the tryouts (see Dick and Carey, 1985; Branson et al., Phase III, 1976). Some argue that ten people should be enough for the first draft and another ten people for the revision. Suppose that you can have 32 soldiers for a period of six weeks to test 100 lessons. Half the lessons require about 30 minutes, the other half average about two hours. You know that the attrition rate is about ten percent a week. You may have replacement soldiers for any that leave the trials, so that you always have 32. At least half of your lessons are dependent on at least one subordinate lesson and a quarter of them are dependent on at least two subordinate lessons. The question is how to arrange the trial to optimize data gathering on all lessons. Try it. You will enjoy playing "what if" games with reality.

Time data. Probably the most critical data to be collected in the trials has to do with time spent on the system. Any cost-effectiveness and cost-benefit analyses will have to use time data as inputs to the cost estimates. Time is a second instance in which course-based and systems-based models vary greatly. Individual professors creating courses do not usually care much if it takes students a few more minutes to complete a course. The students' time has no economic value to the university or the professor. In dramatic contrast, when learners are paid to take instruction as they are in military or industrial settings, efficiencies in time translate directly to dollars.

Regardless of how important the data are, providing for good data collection is not easy. In JSEP, soldiers are taken from their regular jobs and assigned to the education center to take the instruction. Being of lower ranks, their jobs may not be exciting. If they refuel helicopters on the cold Kansas prairie, or if they repair tanks in the Oklahoma summer sunshine, they may not be as interested in leaving the pleasant surroundings of the education center as quickly as they otherwise might. It is much nicer to be in the air conditioned comfort of a clean learning center and play on the computer all day than it is to do what they would be doing if they hurried through the lessons. Soldiers are humans first. Most soldiers will work at a reasonable pace, but since there are few incentives to provide for a fast pace, it is not always possible to get true estimates of optimum completion times.

Summative Evaluation of JOBS

The JOBS program is considerably further along in the evaluation process than JSEP, with a preliminary evaluation of JOBS reported by Baker and Huff (1981), and a more comprehensive report issued by Baker and Hamovitch (1983). The latter report contained extensive data to assess the impact of the JOBS program on the operational problems of the Navy. Preliminary results on the tryouts of the JSEP program have been reported by Branson and Farr (1984) and Peterson and Farr (1984). It will be several years before sufficient data are available to assess the impact of JSEP on Army operations.

The objective of the JOBS evaluation ". . . was to evaluate the job-oriented basic/prerequisite skills training program to determine whether it can compensate for the skill deficiencies of lower aptitude personnel such that they can successfully complete Navy technical schools and perform to standard in the fleet." (Baker and Hamovitch, 1983, p. vii)

Approach. Between 1977 and 1981, about five thousand JOBS-eligible candidates were invited to participate in the training. Of those, some 3000 volunteered and were randomly assigned to one of two treatment groups: the JOBS direct-track, who went to the training immediately following completion of their recruit training (N = 1216), and the JOBS delayed-track who completed apprenticeship training and spent some time in the fleet before participating (N = 1802). A control group of JOBS eligible recruits (N = 2308), school-qualified recruits (N = 1050), and personnel who had spent time in the fleet before attending technical schools (N = 276).

The major findings can be summarized as follows:
1. About 96% of the JOBS qualified personnel who attended the training graduated.
2. The JOBS delayed-track group had a significantly higher number of attrites (those leaving the program) from JOBS school than the JOBS direct-track group.

3. Of those JOBS school graduates on whom data were available (N = 1256), 79% graduated from technical school and 21% attrited. Comparable percentages for school-qualified personnel are 90% and 10% respectively.
4. Thirty-three months after the two groups had graduated from technical school, twice as many of the school-qualified group had been discharged from the fleet. In the summative sense, there was only a 3% net difference in retention between the JOBS group and the school-qualified groups.
5. Twice as many minorities were in the JOBS-eligible group than were in the school-qualified group.

The authors' conclusions were:

> It appears that the JOBS program has the potential for attenuating Navy technical manpower shortages and contributing to minority upward mobility. Considering the significantly lower fleet discharge rate of the JOBS group, the Navy may be unduly constraining its manpower options by excluding these personnel from consideration as eligible for technical training. (Baker and Hamovitch, 1983, pp. vii-viii).

Discussion. When large projects of the type reported by Baker and Hamovitch are carried out by the service laboratories (the JOBS evaluation was done by NPRDC), they often can do extensive follow-up and impact studies that typically cannot be carried out by contractors or investigators in other occupations. Not only can they design and implement studies on short-term bases, they can capture longitudinal data and come ultimately to reasonable conclusions about cost-effectiveness and cost-benefit. While it is of interest to speculate about cost benefit analyses in military settings, in this study the alternative was to have no one in the billet. Since that alternative is not acceptable, the program costs what it costs. Whether it can be redesigned and made more efficient will depend on how well it is managed.

From all data available, it appears that the JOBS program was highly successful. As with any evaluation study in field settings, there are many threats to validity. When the Ns get as high as they were in this study and the results are as important in terms of effect size as these are, threats to validity become less fearful.

What will be the future of the JOBS program? That will depend on many contingencies, including the quality of new accessions, the continuing success of the program, and cost considerations.

JSEP Evaluation

As this chapter is written, JSEP field trials were being conducted at the fourth site, Fort Sill, OK. There, sixteen PLATO terminals were installed for trials that would last about six weeks. Sixteen soldiers were on the terminals in the morning and another sixteen were there in the afternoon. A control group of soldiers who did not receive JSEP during the initial tryout period was used to compare performance. The Fort Sill performance data were not available at press time. However, data from the other three field sites indicated that the soldiers really appreciated the CBI systems, often refusing to take breaks. Their performance on the tryout materials was good, showing positive gains on most variables of interest.

Following the results of the tryouts and using revised materials, a demonstration period for JSEP was planned. This demonstration period was intended to inform others in the Army of the potential use they might have for the program. Army-wide implementation was being delayed pending the availability of the finally configured EIDS delivery device.

Concluding Thoughts

Instructional systems development has been practiced in military settings since the Second World War. The first model, then called the "systems approach to training," was published by Robert B. Miller (Miller, 1954). Since that time there have been literally thousands of applications of the models and procedures that vary considerably in characteristics and appearances. The body of knowledge that is instructional psychology continues to grow at a fast pace in both the equipment use and instructional strategies areas. Regardless of how sophisticated the field becomes, it is putting these models in place and operating them efficiently that remains the most significant challenge.

We have seen that there are many ways to accomplish each of the fundamental phases of the work—from job analysis to evaluation and control. What makes a systems approach a systems approach is the elaboration of goals and the generation of alternative means to achieving those goals. In these two contrasting projects, we saw that the means of instruction varied greatly, while the goals were held constant. We have seen that in applications, compromises and trade-offs must be made. It is simply not possible for ISD projects to subordinate the needs of the organization to the needs of the ISD project—the project must adjust to the needs and mission of the organization. Models must be intelligently applied, not blindly followed.

There is a final, but critical, point to be made in the implementation of ISD projects. In systems-based models and applications, it is essential to success to get the entire system up and operating before becoming overly concerned with the fine-tuning of the courseware. In course-based models, find-tuning of the courseware is the critical element. In putting together a large curriculum, becoming overly concerned with maximizing the performance of individual lessons before the entire system is operating will always lead to a suboptimal solution. In that sense, an instructional system is never completed.

Notes

1. The views in this chapter are those of the author and do not necessarily reflect the view or official policies of the Army Research Institute or the Navy Personnel Research and Development Center.
2. The author wishes to recognize Dr. Beatrice J. Farr of the U.S. Army Research Institute and Dr. Meryl Baker of the U.S. Navy Personnel Research and Development Center for valuable suggestions and for improving the accuracy of the manuscript. Any errors remaining are the sole responsibility of the author. Thank you Bea. Thank you Meryl.

References

Anderson, C.L. (1986). *Historical profile of adult basic education in the U.S. Army.* Unpublished doctoral dissertation, Teachers College, Columbia University, New York.

Andrews, D.H., and Goodson, L. (1980). A comparative analysis of models of instructional design. *Journal of Instructional Development, 3*(4), 2-16.

Back, S.F., and McCombs, B.L. (1984). *Factors critical to the implementation of self-paced instruction: A background review.* Air Force Human Resources Laboratory Technical Report TP 84-24. Lowry Air Force Base, CO (AD A145 143)

Baker, M. (1978). *Determining Rating Clusters of Class "A" Schools.* San Diego: Courseware, Inc.

Baker, M., and Hamovitch, M. (1983, January). *Job oriented basic skills (JOBS) training program: An evaluation.* (NPRDC Tech. Rep. 83-5.) San Diego: Navy Personnel Research and Development Center. (AD A124 150)

Baker, M., and Huff, K. (1981, November). *The evaluation of a job-oriented basic skills training program—Interim report # 1.* (NPRDC Tech. Rep. 82-14.) San Diego: Navy Personnel Research and Development Center. (AD-A107 895)

Branson, R.K. (1979) Implementation issues in instructional systems development: Three case studies. In H.F. O'Neil, Jr. (Ed.), *Procedures for instructional systems development.* New York: Academic Press.

Branson, R.K., and Farr, B.J. (1984). The job skills education program: Issues in design and development. *Proceedings 26th Annual Conference of the Military Testing Association* (pp. 791-796). Munich. (AD B096 442)

Branson, R.K., Rayner, G.T., Cox, J.L. Furman, J.P., King, FJ, and Hannum. W.H. (1975). *Interservice procedures for instructional systems development.* (5 vols.) Ft. Monroe, VA: US Army Training and Doctrine Command. (AD A019 486 through 019 490; ERIC No. ED 122 018 through 122 022).

Dick, W., and Carey, L. (1985). *The systematic design of instruction* (2nd ed.). Glenview, IL: Scott Foresman and Company.

Duffy, T.M. (1985, Spring). Literacy instruction in the military. *Armed Forces & Society, 11*(3), 437-467.

Farr, B.J. (1986). *Improving job skills education for soldiers.* Paper read at the Annual meeting of the American Educational Research Association, San Francisco, April.

Frederiksen, J.R. (1986). A cognitive task analysis approach to basic skills learning. Session 11.28 of the Annual Meeting of the American Educational Research Association, San Francisco, April.

Gagne, R.M. (1971). Paper presentation at a formal dinner commemorating his election to the presidency of the American Educational Research Association. Florida State University, Tallahassee.

Harding, S.R., Mogford, B., Melching, W.H., and Showel, M. (1981). *The development of four job-oriented basic skills (JOBS) programs.* (NPRDC TR 81-24.) San Diego: Navy Personnel Research and Development Center. (AD-A106 370)

Miller, R.B. (1954). *Some working concepts of systems analysis.* Pittsburgh: American Institutes of Research.

Peterson, G.W., and Farr, B.J. (1984). The job skills education program: Results from preliminary tryouts. *Proceedings 26th Annual Conference of the Military Testing Association* (pp. 797-802). Munich. (AD B096 442)

Sticht, T.G. (1975). *A program of Army functional job reading training: Development, implementation, and delivery system.* (HumRRO Tech. Report) Alexandria, VA: Human Resources Research Organization. (AD A012 272)

U.S. General Accounting Office. (1983). *Report to the secretary of the Army.* GAO/FPCD-83-19, June 20.

Wilson, L.S. (1986). Improvements in military training: 1976-1986. Session 41.07 of the Annual Meeting of the American Educational Research Association, San Francisco, April.

Appendix B

Instructional Systems Development in a Large Governmental Agency

Martha Brooke
Internal Revenue Service

Introduction

Training program development in a large government agency is a study of coordination—coordination of communication and coordination of resources. Geographically spread out and organizationally complex, a large federal agency must make its decisions and requests through channels to assure thorough and appropriate communication. These channels, for any one communication, may involve not only many levels but also a number of branches within hierarchies.

This case study, which involved the Internal Revenue Service, follows a major program redesign in management training. The effort was sustained over a seven-year period and encompassed numerous personnel changes and a major technology revolution. The IRS has approximately 110,000 employees. It is divided into major organizational areas according to work function. Six major functional areas are: Examination, Collection, Taxpayer Service, Criminal Investigation, Resources Management, and Employee Plans and Exempt Organizations. Functions can be thought of as having two major organizational arenas—"the National Office" and "the field." The National Office, located in Washington, DC, sets policy and program objectives for a function and the field implements them. National Office does not specify how the field uses field resources (budget and personnel) in their implementation. In addition to ten large service centers where tax returns are processed, the IRS field organization has over 300 "outlying" posts of duty that are contained within 63 districts. The service centers and districts, in turn, are organized into seven geographical regions (North Atlantic, Mid-Atlantic, Southeast, Central, Midwest, Southwest, and Western) of the country.

In the IRS, there are two types of executives: (1) *functional* such as Assistant Commissioners (AC's) and Assistant Regional Commissioners (ARC's) who respectively direct their function's National Office and field operations, and (2) *cross-functional*, such as Regional Commissioners (RC's), District Directors (DD's) and Service Center Directors (SCDs) who direct major organizational components containing a number of functions.

Below the rank of the executives and their assistants, there are three levels of functional management: top, middle and entry level. Top managers are the division chiefs and their assistants. Middle managers are managers of managers and can be either branch or section chiefs, depending on whether a section has subunits or not.

There is not a simple hierarchical line of authority even within the major work functions of the organization. At National Office, a function's Assistant Commissioner's office sets policy for that function; and yet, functional activities in the field are located in organizational components directed by cross-functional executives. The coordination of the multiple hierarchies and levels for a training development project is accomplished by using a training project agreement. At the initiation of a project, a contract called "project agreement" is negotiated at National Office by the assigned training program manager with the client function's Assistant Commissioner office.

The training project agreement exists within an environment where training development projects are managed by a separate training staff, with the function as client. The functions provide the technical expertise for the development and review of training products and for classroom instruction.

The players on a training development team (i.e., the training designer, the content specialist[s], and manager) can be characterized in the following manner on a typical IRS program development effort resulting in a product to be used service-wide. The

ISD in a Large Governmental Agency

training designer is the program or course manager from the National Office's Training and Development Division.

The content specialists are members of task forces from the client function. IRS makes extensive use of task forces for program development. Small groups of technical experts, consisting of as few as two to as many as 14 individuals, may be convened to provide functional input for program design, development and/or evaluation/revision.

The training team manager's role is played by two management hierarchies: training's and the client function's. The client function's training coordinator serves as its coordinating representative and interfaces with the training designer. In drawing up a project agreement, typically the training program or course manager presents a draft plan that spells out estimated activities and resources to the functional training coordinator. Any negotiated changes that those individuals and/or their management deem necessary are made before a final agreement is circulated for signature.

Project agreements for service-wide programs that are developed at National Office, even though they use input of field task force members, need only be negotiated and signed within the National Office, up to the level of the Director of the Training and Development Division and the Assistant Commissioner of the function. For field course development efforts, which are programs managed by National Office and developed in the field, the project agreement is negotiated among the training and client function offices of both National Office and the hosting region.

Project agreements are logistically important. Once a project agreement for either National Office or field course development has been signed at all levels, requests for functional expertise (i.e., nominations of task force members and classroom instructors for a pilot) can be made directly to the field. Field training and function staffs find the appropriate talent within the region and convey the nomination(s) to the program or project manager. This saves the tedium of having to go through all training and functional levels at both national and regional offices for every activity requiring a coordinated effort during the life of a project.

The rigors of developing a project agreement force the program manager to plan from the project's onset and negotiate a project's travel and per diem budget, staff year obligations and schedule. A training program development project agreement may cover only the initial analysis activities, with later agreements following after Training and Development has had an opportunity to give its curriculum report. Or, a project agreement may cover all five phases of program development, calling for extensive and long-running functional input over several task forces and a pilot class. In this case study, there are two kinds of program development within the areas of management training—a macro and a micro.

The Macro Program

The macro program, the redesign of management training for all IRS middle managers, focuses on the analysis and design phases of the development process.

Front-end Analysis
In 1978, the Civil Service Reform Act was passed; this affected requirements of management training. In 1979, the Management Programs Section within the Training and Development Division at National Office chartered an executive advisory group to provide high-level executive recommendations for its multi-functional work. The group, called the Management Training Advisory Committee (MTAC), recommended that the Management Programs Section conduct a needs assessment for all three levels of management. While pre-1980 front-end analysis efforts focused on common needs for training of all IRS mid-level managers, the 1980 analysis changed this focus by posing two major questions. First, were the training needs of entering middle management the same as those in top management? Second, were there substantial functionally-tied training needs beyond those that could be addressed cross-functionally?

The method of the analysis involved three steps: the use of task forces; validation of task force work by executives and upper management; and the analysis of the task forces' work by the training and development team. The training and development team within the Management Programs Section consisted of the mid-level program manager who was assisted by an organizational development specialist, the program manager of the entry-level management training program, and another Employee Development Specialist who was working with the mid-level program.

Small task forces were convened for four major functional areas: Examination, Collection, Resources Management and National Office. (Between Examination and Collection, approximately 70% of all IRS mid-level managers are represented.) The Examination and Collection Task forces were comprised of middle and top managers; the other two, only middle managers.

To assure comprehensive content coverage, in addition to the task forces for the two largest functions, questionnaires were sent to 101 Examination and Collection branch and division chiefs. The return rate for the questionnaires was 80%. Both task force members and the respondents to the task analysis questionnaire were given a list of 29 task statements as a beginning point for their analysis. This list had been a product of the 1975 redesign effort.

At the IRS, training programs are validated by subjecting them to additional technical and higher management review. In this project, an Assistant Regional Commissioner and an Assistant Commissioner Designate validated the Examination and the Collection task forces' work. The Resources Management task force's work was reviewed by a Division Chief and an assistant to a Regional Commissioner. The product of the National Office task force was sent to all Assistant Commissioners for validation.

The Training and Development team for this project analyzed the lists of task, subtasks, knowledges and skills generated by the task forces and the returned questionnaires. Using the 1975 task list as a point of comparison, the team consolidated the task data and determined component knowledges and skills.

Macro Design

The design phase was conducted by the same Training and Development team that conducted the front-end analysis. The result of their analysis reflected the nature of its inquiry approach. Findings confirmed the need for functional as well as cross-functional training, and determined that mid-level management training, in order to accommodate the needs of an evolving career, should be more than a one-time offering; also that mid-level training should be extended to include all managers of managers, including those in positions below the GS-12 grade level. The redesign basically called for three types of mid-level management training (see Figure 1: Timeline Relationships of Functional, Cross-Functional and Continuing Professional Education Training).

Functional Training

In the redesign, the functional training would have two parts—an On the Job Training (OJT) component and a classroom seminar. It would address "how to manage work" and incorporated instruction on the role of the middle manager in special emphasis programs.

The OJT component would address immediate needs of newly appointed mid-level managers and put those managers in a coaching relationship with their direct supervisors. This would provide the new mid-level managers with necessary instruction while they wait for a classroom seminar to be convened.

A couple of times a year, a one-week classroom seminar would be given for newly appointed mid-managers within a given functional area. These seminars would be hosted by a regional office, but would have managers from all over the country as participants. The seminar would expand upon the OJT topics and address others that required large-group participation. IRS classroom instruction usually would involve 15-24 participants and 3-4 instructors.

Cross-Functional Training

The cross-functional training, usually hosted by a regional office, would address the "how to manage people" aspect of management training over a two-week period. These classes would be held eight to ten times a year because of their large audience of newly selected mid-managers from all functions, from all over the country.

Continuing Professional Education

Continuing Professional Education (CPE) would involve three types of training: update, gap and depth training. These would be

ISD in a Large Governmental Agency 261

Figure 1 Timeline Relationships of Functional, Cross-Functional and Continuing Professional Education Training.

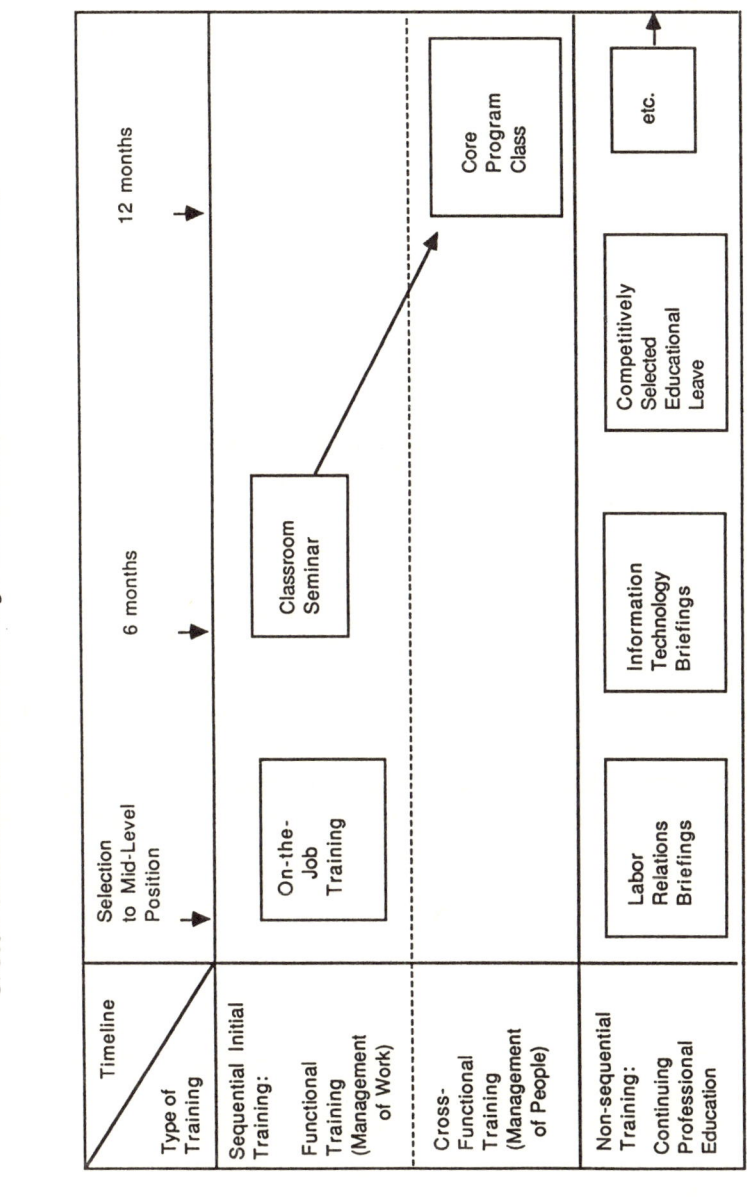

Timeline / Type of Training		6 months	12 months	
	Selection to Mid-Level Position			
Sequential Initial Training: Functional Training (Management of Work)	On-the-Job Training	Classroom Seminar		
Cross-Functional Training (Management of People)			Core Program Class	etc.
Non-sequential Training: Continuing Professional Education	Labor Relations Briefings	Information Technology Briefings	Competitively Selected Educational Leave	

designed to meet the needs of managers who complete their careers at the mid-level. (The formal training for these managers, who may spend decades in mid-level management, had formerly consisted of the two-and-a-half-week course attended upon entering mid-level.) Recommended CPE avenues for this audience would include IRS self-instructional mini-courses and seminars, a competitive educational leave of absence policy and an incentive system for study outside IRS.

Scheduling

The training and development team for this project submitted its proposal for the redesign to the Management Training Advisory Committee in August 1980. A summary of the proposal was then circulated by the Director of Training and Development to all the Assistant and Regional Commissioners in IRS for their review. This summary contained overview charts showing the major tasks to be covered in cross-functional training and in the functional training of each function that participated in the analysis. A timetable was also proposed.

The timetable called for immediate design and development of the cross-functional "people" training. The other micro efforts, the development of the functional training programs, however, were staggered. Functional programs would more heavily involve limited IRS resources, especially training development resources. The Service Centers, nevertheless, did have the resources to begin developing functional training for their section chiefs; and so, theirs was targeted as the functional program's prototype.

The beginning of the functional efforts for Examination and Collection followed in a year, and six months later those for Resources Management and National Office began. Each developmental effort, it was estimated, would take about a year to complete. The course needs-analyses and development of remaining functions (e.g., Taxpayer Service and Criminal Investigation) were to be scheduled later.

The macro program ended with the acceptance of the mid-level redesign program. Subsequent project agreements were necessary for the development of the micro programs. The following is a short update of the execution of the overall mid-level training plan.

ISD in a Large Governmental Agency 263

Immediately following the executive review and acceptance of the macro mid-level design, development of the cross-functional "people" course began. Aspects of its design were fed into the ongoing (1975 design) course, until they completely subsumed that course. The design, using the competitive contracting route, engaged consultants for approximately 20 courses conducted over a two-year period.

The original 1980 redesign had also called for the functional OJT and classroom seminar to be taken within the first six months of a mid-level manager's appointment. The "people" course followed by between six months and a year. This timing was changed in order to get useful questionnaire assessments of the cross-functional training. Prior to, and one year following the "people" course, not only did the program participant fill out questionnaires but also the participant's direct supervisor and subordinate managers responded to questionnaires regarding the participant's management style. In order for the participants to be known in their new position by these other respondents, attendance of this people-skill training was rescheduled to a year after they had assumed a mid-level position.

The Micro Program

The work I was involved with was a "micro" project within the "macro." The micro level was the development of a functional "how to manage work" mid-level management curriculum. I joined the IRS and the staff of its National Office Management Programs Section in November of 1983. As National Office Program Manager, I was assigned the development of the Collection functional mid-level program. At that point, only two other courses had been developed as part of the larger curriculum design. The cross-functional "how to manage people" course had been in place several years and one functional "how to manage work" course for the Service Center Section Chiefs had had its first conduct.

When I joined the effort, one other EDS was starting course designs: I was working on Collection's; the other EDS, on Examination's. The other EDS had been the field project manager for the Service Center prototype and served as my coach for IRS procedures. By January 1984, we had each negotiated project agreements with our respective function client. The agreements were for National Office development, with nation-wide representation on all task forces and questionnaire respondent samples. The project agreements were going through channels for signature.

Later that month, however, the IRS Executive Committee, consisting of all Assistant and Regional Commissioners, met in Washington. After being briefed on the mid-level management training program, five Regional Commissioners decided to facilitate the development of the functional curriculum by sponsoring one regional development each. The commissioners also expressed an eagerness to have the courses piloted by the end of the 1984 calendar year. Western Region volunteered to host the field course development of the Collection function's course.

Front-end Analysis

Even though the 1980 redesign of mid-level management training contained an analysis of the functional training needs for Collection, the course-level analysis was to be more extensive. Because this project was now a field course development effort, as National Office Program Manager, I had to negotiate a new project agreement not only at National Office but also with the host Western Region's Offices of the Assistant Regional Commissioners of both Collection and Resources Management. A draft was finalized and started through regional and national office channels for signature. The major activities of this project proposal were: a task analysis; three task forces, each followed by an executive review when appropriate; and a pilot of the classroom seminar.

The Task Analysis consisted of updating the 1980 task analysis of Collection management (April 1984).

The three Task Forces (Course Analysis, Design and Development) were for:

(1) Assembling a two-week task force of seven experienced mid-level managers, one from each region (May 1984). This task force would:

—review the updated task analysis information
—specify objectives and criterion measures

—determine the appropriate training method for each objective (i.e., OJT, classroom seminar or Continuing Professional Education module)

The results of this task force would be reviewed by the Regional Commissioners and an Executive Review Committee.

(2) Assembling a three-week task force which would also be comprised of seven experienced mid-level managers, one from each region (June-July 1984); this task force would begin development of the training materials. If the product included OJT materials that needed to be piloted before the classroom seminar, the materials would be reviewed by National Office and (Western) Regional Office Collection and Training. Subsequent to this review and revision, the materials would be submitted for executive review.

(3) Assembling another three-week task force, again consisting of seven experienced mid-level managers, one from each region. This task force would continue developing training materials. The materials would be reviewed and validated by National Office and (Western) Regional Office Collection and Training. The materials and any revisions, then, would be sent for executive review.

With the pilot of the classroom seminar (December 1984), the target date for the availability of final-form materials was spring 1985. The project agreement was drafted and sent for signature in early March 1984; it was fully signed by mid-April.

The list of Collection functional tasks presented as part of the 1980 redesign of mid-level management training was admittedly an overview, not a detailed representation. Therefore, for the Collection course, there were three additional formal sources of job/task analysis. The first two, a function-wide and a service-wide survey, fed into a third, a task force.

There was also an informal fourth source: telephone interviews that we conducted with two instructors who had recently completed the first full conduct of the Service Center Section Chiefs' training program.

For the remainder of this course development description, I will use the term "course manager" to depict the role which I and my successor, as the National Office Program Manager, played in conjunction with the Western Region Project Manager. I left the project after the executive reviews following Task Force II.

Function-wide Survey

When the project agreement was being negotiated in Western Region, the Assistant Regional Commissioner for Collection indicated he wanted Collection-wide training needs input for the task forces hosted in his region. Therefore, the course manager designed a function-wide, open-end question survey to gather input for task forces to analyze, design and develop this program. A questionnaire form was tailored for each of the seven types of Collection mid-level manager. Each questionnaire asked:

- What are the primary duties/tasks of a Collection mid-level manager?
- What do mid-level managers do that is different from what first-level managers do?
- What topics related to work management should be included in mid-level training?
- How has the Automated Collection System changed the job of the Collection mid-level manager?
- What authorities and responsibilities differentiate the work of top managers from that of mid-level managers? (For top management respondents—division chiefs and their assistants.)

The field course development project agreement had to be approved before the needs survey could be sent for review by the Assistant Commissioner for Collection. The agreement was not signed until mid-April. Because of the close (mid-May) response date on the survey, the Assistant Commissioner's Office refused to sign it. Subsequently, a redated survey was sent to the Assistant Regional Commissioner (ARC) for Resources Management in each of the seven regions, with information copies to the ARCs-Collection. The ARC-Collection offices then had to decide what constituted an appropriate regional response to the questionnaire for each of the seven mid-level positions. This illustrates how National Office states "what" needs to be done, whereas field decides "who" and "how" a program objective is met using field resources.

The delay caused the responses of the finally circulated survey to arrive in the middle of the second task force, not providing input for the first task force, which focused on needs analysis. Consequently, the responses to the survey did not undergo any organized analysis; rather, they were studied "to confirm or augment the first task force's work."

Service-wide Survey

The management of the National Office Training and Development Division was eager to get updated, service-wide input on "common core" issues. These were issues that would have to be covered to a specified minimum in all the mid-level functional efforts.

To accomplish this, on March 1, 1984, a memorandum was sent from the Director of Training and Development to all Assistant and Regional Commissioners, asking their input and stating that their responses would be shared with the Management Training Advisory Committee (MTAC) on March 28. At that time, a decision on the necessary topics would be made.

Responses to the memorandum were initially organized by one of the four mid-level program managers. This initial organization was used as a basis for group work by all mid-level program managers and the Management Programs Section Chief. Their consolidated effort was then shared with MTAC, resulting in a list of ten essential topics, which we designated as common issues of mid-level managers to be addressed in functional management training courses:

    Office automation/Information technology
    Fiscal responsibilities and the budget process
    Operational reviews
    Planning
    Training and development of employees
    Meetings
    Delegation
    Position management
    Organizational relationships
    Security, disclosure and integrity

This survey resulted in a list of topics that were generated through high-level executive input and adopted by a cross-functional executive committee with responsibilities for the training program. The ten topics were incorporated into the curriculum of each functional effort by task forces.

Task Forces I, II and III

As stated earlier, the objective of Collection's first task force was to: review function-wide survey, specify objectives and criterion measures, and determine the appropriate training method for each objective (i.e., OJT, classroom seminar or CPE module). Since the "review" of the function-wide survey did not occur until the responses were received during the second task force, the first task force had to generate its own job/training needs analysis, keeping in mind the common core issues and using some other sources of input.

Task Force I was comprised of seven Collection mid-level managers. Five were representatives of the initial mid-level positions; two were representatives of top management, supervisors of the initial positions:

2 Collection Division Chiefs (Top Management)
3 Collection Field Branch Chiefs
1 Automated Collection System (ACS) Branch Chief
1 Service Center Collection Branch Chief

A modified nominal group method was used to generate the major duties and tasks of the mid-level functional management training. Only after the task force had had this opportunity to do its own thinking did the members look at the service wide common core requirements, topics covered in the training for first-line supervisors and topics covered by the "how-to-manage-people" cross-functional course for mid-level managers.

Other sources introduced at this time and made available to the task force members included: relevant sections of the Internal Revenue Manual (IRM); the task analysis lists and charts from the 1980 redesign; materials from the Service Center Section Chief's prototype; and the products of just-completed task analysis task forces for two other functional efforts.

The specified tasks became the training goals of the course. At this point, we moved the activities of Task Force I from the course analysis phase to the design phase.

The Course Design Phase

The design work of Task Force I included writing behavioral objectives for each task, developing a content outline (subordinate skills and knowledges) for each task, sequencing the instructional order of the major "duties" or topic areas under which the tasks were organized, and specifying teaching methodology (i.e., OJT, classroom seminar or CPE module) for each topic.

Cryptically stated, the major course topics and their subordinate goals addressed the following areas:

Transition to Mid-level Management
    New role as a mid-level manager
    Delegations
    Time management

Goal Setting
    Organizational objectives
    Merit pay

Communications
    Communication through meetings
    Functional communications
    Cross-functional and outside-organization communications
    Briefings

Office Automation/Information Technology
    Initiating office automation programs

Problem Resolution and Other Sensitive Taxpayer Situations
    Problem resolution program
    Controlled correspondence
    Taxpayer inquiries

Resources
    Human resources
    Property

Administrative Duties
    Fiscal responsibilities
    Organizational safeguards
    Labor-management relations

Equal Employment Opportunity
    Equal employment opportunity program
    Special emphasis programs
    EEO complaints

Managing subordinate managers
    Development of subordinates
    Corrective actions

Reviews
    Operational reviews
    Program reviews

The task force was then divided into groups of two or three persons each. Each group was given one or two topics and told to analyze its subordinate tasks into prerequisite skills. The task force had difficulty organizing the subtasks using a hierarchical "organizational" chart system. The more traditional outline approach was used.

Objectives—Instructional Outcomes
Initially, the task force members wanted to think in terms of what the incumbent should be able to do on the job. In order to set the task force members thinking in terms of instructional outcomes, the course manager could not even use the old instructional designer's fallback position: if they cannot think in terms of "objectives," go for the test. These mid-level managers did not like the idea of managers taking tests—that occurred in lower-level technical position training.

The key for getting the task force members to think in terms of instructional outcomes lay in the traditional format of all IRS recruit training. Recruit training usually has the following format for every lesson: objectives, information presentation, practice exercises with feedback, and test (usually given over a number of lessons at a later time). The course manager asked the task force to describe the practice exercise that would follow a lesson's information presentation. This would give the same outcome as a test situation. The objectives were first worked on in small groups of two or three and then further developed by the task force as a whole.

After specifying the objectives, individual task force members developed detailed content outlines for assigned lessons. These outlines were subsequently reviewed and accepted by the whole group.

Selection of Training Method

A turning point in the design of this course came when the major topics and their lessons were sequenced, and a decision regarding training method (i.e., OJT, classroom seminar or CPE module, according to the macro design) was made.

Throughout the two weeks, the task force members repeatedly said "Reviews are everything." Reviews were written by the mid-level manager assessing the performance of a subordinate manager and the effectiveness of specific programs. At this sequencing point in the development process, the course manager was working with the task force to see if the "review" lesson would provide the course with a start-to-finish process around which the entire course could revolve. The review lesson just did not work, because goal setting in the first lesson involved more than manager and program reviews. Then a task force member saw that the big assessment or review was "branch assessment," which does account for all the variables addressed in the goals lesson. In fact, it accounted for the whole course!

The entire course was conceptualized as large "in-basket" exercise, an adaptation of the management assessment tool. Each course participant would play the same role—that of manager of fictitious Branch A. At the beginning of the course, the participants were given an organizational chart of their branch and short biographical sketches of their superior manager and their subordinate managers. Such descriptions, of course, contained seeds of many of the management issues that the participants would have to deal with as the course unfolded.

In the first of nine lessons, the participants set goals for their branch and for their own Merit Pay expectations. (See Figure 2: The Piloted Course Design for Collection Mid-level Functional Management Training.) The succeeding lessons had exercises in-

272   *Instructional Systems Development in Large Organizations*

Figure 2 The Piloted Course Design for Collection Mid-Level Functional Management Training.

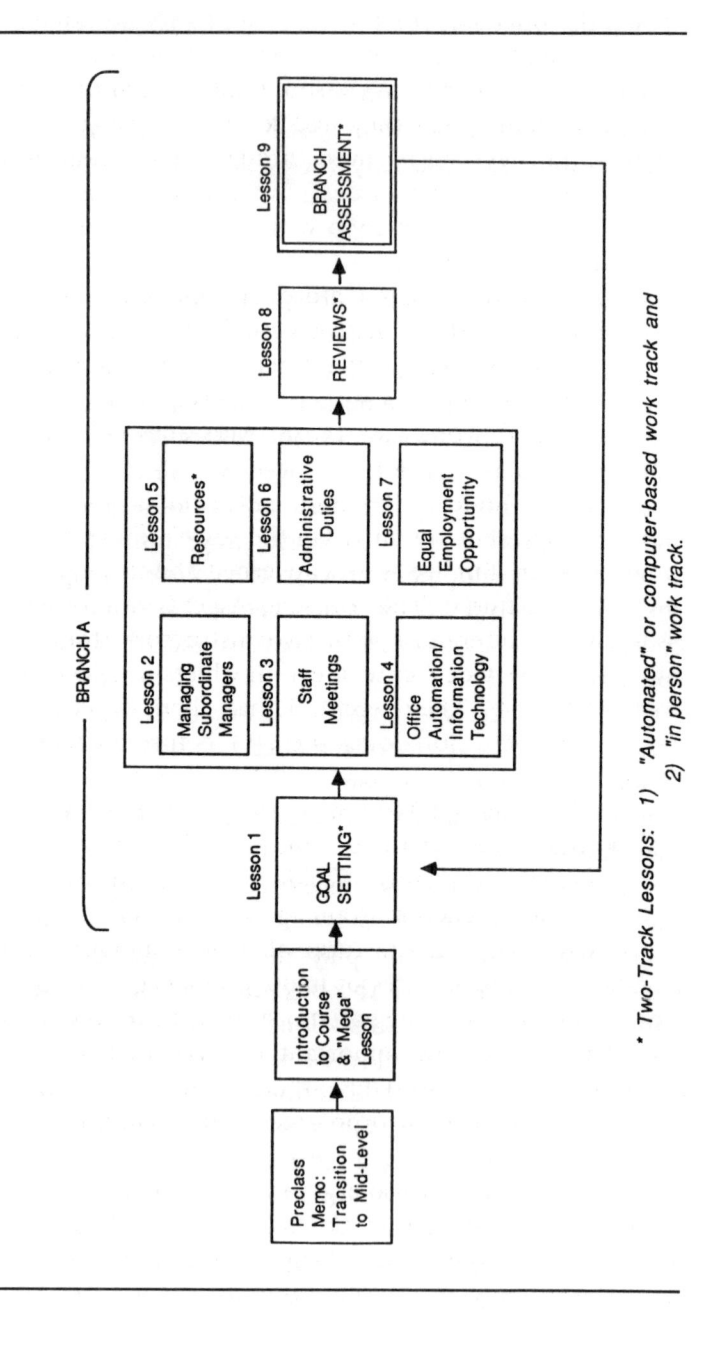

* Two-Track Lessons: 1) "Automated" or computer-based work track and 2) "in person" work track.

volving the operation of their branch. In the eighth lesson participants, as individuals, wrote a review of one of the subordinate managers; and, in the final lesson, they assessed the branch, taking into account the goals they had set in the first lesson and all the information they had received throughout the course. In this conceptualization, the overall design of the course was that of a process, having a continuous loop with necessary goal revision following branch assessment.

Another major design feature of this course was having two tracks for the exercises of three crucial lessons: Goals, Resources and Reviews, as well as the final Branch Assessment. The two tracks were for those managers involved with "automated" or computer-based work and those overseeing work done "in person."

The 1980 Mid-level Redesign had specified two types of functional management training: On the Job Training (OJT) and Classroom Seminar. But we were concerned about the implementation of the OJT material. The course manager had informally conducted telephone interviews with two instructors from the first conduct of the Service Center Section Chiefs Mid-level Management Training Course. They reported that about three-fourths of the participants were not doing the OJT before coming to the seminar.

What had been OJT topics for the Service Center's course were pivotal materials for our mega- (in-basket) lesson simulation. Our task force, therefore, decided that in lieu of extensive OJT material, the Collection training program should provide some pre-class materials in the form of a memorandum to be sent to both selector and selectee of mid-level management positions. The memorandum would suggest transitional issues for mutual discussion. The bulk of the important material that had been in OJT, according to the 1980 redesign, and as used in the Service Center prototype, would be incorporated in the branch assessment model taught in the seminar.

The task force was also eager for this program to be implemented in a fashion similar to the cross-functional "people" course—off-site, away from an IRS facility, at a hotel/conference center. An off-site location would promote evening, informal sessions among the participants and with guest executives. A similar implementation format would also use an executive instructor in addition to the regular instructors. The executive instructor could speak about the importance of organizational relationships and convey topic concerns to guest executives for their evening sessions.

This course model provided a mechanism to assess student performance; the next-to-last lesson (Reviews) called for the participants to write up an evaluation of one of their subordinate managers. This exercise involved synthesizing materials from the previous lessons. The instructors of the course read each review to determine the participant's mastery of the material.

The specification of the media (e.g., instructor, videotape or guest expert) for the presentation of instruction as well as the mode of the practice exercises (i.e., individual, tutorial, small or large group format) were not worked out until the second task force. In general, this course still preserved the classroom format for the delivery of information. It was in the practice exercises that the mega-lesson "continuous example" was used.

Task Force II

The first part of Task Force II was devoted to drawing up the simulated branch model. This included the cast of characters and the placement of simulation problems within that cast and the branch's operations (e.g., who plays golf with the boss, who has asked a subordinate for a date).

Review of the Course Design (Formative Evaluation)

As called for in the project agreement, the course's design received not only the traditional training and functional reviews at the national and regional offices, but it also underwent extensive and service-wide review. After the five course development efforts were begun in the spring, the management of National Office Training and Development was eager to provide more executive involvement for each. This involvement would be more functionally and program specific than that provided by the on-going Management Training Advisory Committee (MTAC).

Five Executive Review Committees (ERCs) were established through a memorandum from the IRS's Deputy Commissioners—one committee for each function's effort. Each committee consisted of four executives and examined the programs at three stages of development: after the objectives and the design had been designated; after the materials had been developed; and after the pilot had been conducted. Of the committee members, the Assistant Commissioner and the Assistant Regional Commissioner were to review the program from a functional perspective; and the District Director and the Service Center Director were to review it from a multifunctional perspective. The committees were not necessarily to convene for each review. Materials could be distributed and conference calls used.

At the same time the course objectives and outline were being sent to the Assistant and Regional Commissioners as feedback to their responses to the service-wide survey, the objectives and outline were also sent to each member of Executive Review Committee for Collection's mid-level functional program. This was in preparation for the committee's first meeting, held a couple of weeks before the third task force.

Like Task Force I, the committee wanted an implementation format similar to the cross-functional course (e.g., conducting meetings off-site and including an executive instructor). The committee felt that designating an executive instructor would signify the importance of the training, and the executive could provide the participants with career counseling. The committee also wanted top-notch instructors with varied backgrounds in Collection chosen for the course's cadre of instructors.

The Assistant Commissioner designate, chairman of the ERC, committed Collection to providing the necessary personnel for Task Force III. There had been difficulties in obtaining the mix of experience desired for the task force. There were only ten Service Centers and 21 Automated Collection System call sites; mid-level managers from those settings had been more difficult to secure for the task forces than the more numerous field branch managers.

The committee wanted at least one of its members to attend part of the pilot and felt its future role was one of oversight of the process of training materials development. Copies of the training materials completed in Task Force III were sent to each member of the ERC, as well as to National Office Collection, Western Region Offices of Collection and Training and other regions for functional review. In January 1985, the Committee reviewed the results of the December pilot.

A meeting of the Management Training Advisory Committee was called for the week before Task Force III began work in Mid-August 1984. At the meeting, MTAC made a general recommendation that all five functional programs have an executive instructor, perhaps an Assistant Regional Commissioner from the function. The committee was also interested in having the conducts implemented in a fashion similar to the cross-functional program (e.g., off-site to promote evening, informal sessions).

The Course Development Phase

The development phase had begun with Task Force II, in late June 1984. Four of the seven members of Task Force I had returned; however, the second task force lacked continuous representation from the Automated Collection System, and totally lacked Service Center representation.

For the first half of the three-week task force, the Second Task Force's members finished the design of the mega-lesson, drawing up the cast of characters and the organization of the fictitious branch. They also placed simulation problems within the cast and the branch's operations. Throughout the planning and writing of the course materials, task force members drew upon their own experiences and "sanitized" real cases in an attempt to make Branch A as realistic as possible.

At this point, decisions were made regarding media (e.g., instructor, videotape or guest expert) and presentation mode (i.e., individual, tutorial or small/large group). Actual writing did not begin until the second half of the task force's three weeks. When the writing began, each task force member was individually assigned one or two lessons to produce.

The materials were continuously reviewed by the course manager as they were being drafted. Each of the task force members also shared their materials with at least two other task force members; thus the materials received an immediate informal functional review. The course materials developed by Task Forces II and III consisted of three volumes: an instructor guide, a coursebook for students and a branch resource book, containing the "in-basket" items such as division memorandums, workload and scheduling guidelines and monthly reports.

Formative Review

As called for in the project agreement, formal formative review of the program's materials, as drafted by Task Forces II and III, consisted of Training and Collection reviews at both National Office and Western Region. As requested by the Collection effort's Executive Review Committee, the materials were distributed for review to the ARC-Collection offices in the remaining six regions. Copies were also sent by the Regional Commissioner of Western Region to each member of Collection's Executive Review Committee. All feedback was collected by the course manager. Appropriate revisions were made before the pilot class.

At this point, the time projections for the project were proceeding on schedule, as estimated in the project agreement. Costs were somewhat higher because of extensive National Office program manager involvement with the task forces, and post-task force editing.

The Course Implementation Phase

The project agreement spelled out a number of "givens" regarding the implementation of this program. Most of these were carried out; several were modified. The agreement called for a pilot in early December 1984, involving 15 participants and three instructors, including a lead instructor. Traditionally, IRS training classes use one instructor for every four to six trainees. The lead instructor takes a reduced lesson-teaching load because of administrative duties. The pilot, however, used four instructors. This was to enable every group to have an instructor available as facilitator during small group activities.

The course design also added an executive instructor. The executive instructor does not necessarily assume lesson-teaching duties, but acts as a resource, providing national perspective on such issues as national budget cutbacks, career opportunities, moving expenses and the Equal Employment Opportunity program. The executive instructor also serves as coordinator for other guest executives, conveying to the guests, several days before their visits, questions raised by the participants.

Although the 1980 redesign called for on-the-job training and a one-week seminar for the teaching of functional "how-to-manage-work" training, this seminar design for Collection incorporated the majority of the prototype's OJT material into the seminar format, thus lengthening the classroom instruction to one and a half weeks and eliminating the OJT phase.

Some other agreement projections remained the same: The estimated life of this course was 2-3 years before revision. With over 300 mid-level managers in the Collection function, the estimated number of trainees per year was 30-45. Two classes would be conducted, with 15-25 attendees each. Trainees were to be newly selected managers of managers, receiving this training within the first year of their mid-level appointment.

The executive instructor for the pilot and subsequent conducts was from the region hosting the class, and had a background in the functional area of Collection. For the pilot, Western Region provided an Assistant District Director.

In early October, the Director of the Training and Development Division sent a memorandum to all Assistant Regional Commissioners of Resources Management with an information copy to ARC's Collection. This memorandum requested nominations for both pilot class participants and the instructor cadre.

For the pilot, it was advantageous to have adequate representation from the major categories of Collection mid-level managers. In order to attain the appropriate mix of field, Service Center and Automation Collection System managers, the Director designated the types of managers each of the seven regions was to nominate.

The Executive Review Committee was eager to recruit a top-notch instructor cadre; therefore, the Director's memorandum requested all regions to nominate instructors with strong Collection backgrounds and previous teaching experience. Each region was asked to nominate one division chief, two field branch chiefs, and one ACS branch chief. The ERC was to select instructors for the pilot class, keeping the remaining list for future classes. Managers who had worked on the development of this course as task force members were not considered for selection as instructors for the pilot.

The requirement of "previous teaching experience" is not a casual requirement at IRS. All classroom instructors at the IRS are required to pass Basic Instructor Training (BIT) before they are allowed to teach a class. This is a nine-day course given by Training, and participants must meet set standards. BIT is considered one of the most difficult courses offered at IRS. After completing BIT, instructors must successfully teach one course during which they are extensively evaluated by Training personnel before they are placed in the instructor cadre. Experience as a successful instructor is desirable for career advancement. Evaluations of instructors are kept on file by regional Training; these include evaluations of their performance in the BIT course and in classes they teach.

All instructors are evaluated every time they teach. An authorized person from Training normally observes portions of a class to assess the instructors and the class conduct according to training criteria. Functional personnel assigned to the training staff evaluate instructors on their technical competence. Such rigor is supported by the functions, who want their trainees to receive the best possible instruction.

Instructor Preparation

Before a major class at the IRS, instructors are given one to two weeks to prepare their lessons and to build a team. This is done at a district or regional training site, away from the work setting. The training personnel, in the case of this pilot, was the course manager who was available to work with the instructors during "prep."

For this program, and for most IRS courses, materials that are piloted include an instructor's guide. This explains the exclusion of task force members from the pilot's instructor team.

For the pilot, prep took place at Western Region's training office in San Francisco between November 26 and December 2; the class was held at a hotel in Oakland over December 2-December 11. There were four instructors: the lead, a division chief;

two assistant chiefs; and a field branch chief. One of the assistant division chiefs had had experience setting up an Automated Collection System site, as an ACS Project Manager.

Pilot Evaluation and Revision

The course manager observed each lesson as it was taught and then collected rating sheets from each participant; he or she also interviewed the instructor and collected marked-up copies of the instructor's guide. Participants rated lesson objectives as to (1) relevance to the job of a Collection mid-level manager and (2) the accomplishment of the objective in the class. A five-point scale, with five being the highest rating, was used. The lesson was also given a similar five-point scale overall rating, and space was provided for comments and suggestions. The course manager consolidated the information on the spot, eventually creating a split column format showing comments with actions recommended. This documentation was used with the Executive Review Committee.

The Executive Review Committee met in Mid-January, 1985. In addition to the course manager, instructors from the pilot, including the executive instructor, attended. The regular instructor team remained at National Office for two additional days in order to write accepted revisions to the course material.

Major Results and Revisions

The end of course evaluation showed a 4.0 rating out of a possible 5.0 for the course overall. The lessons considered most beneficial were: reviews, resources, branch assessment, and managing subordinate managers. These were the most fundamental of the lessons, involving extensive interlesson coordination.

The greatest shortcoming of the course was insufficient material for the computer-based track. For the most crucial lessons, there had been two tracks: one for managers who supervised "in-person" work, and another for those supervising computer-based Collection activities. Computer-oriented managers were those from the Service Centers and the Automated Collection System call sites. Because of their scarcity and the tremendous program demands in those areas at the time of this training program's task forces, Task Forces II and III were under-represented by ACS and the Service Centers.

To make up for this deficiency, the course manager traveled to the ACS call site where one of the pilot instructors was an Assistant Division Chief and the ACS Branch Chief had been a class participant. In consultation with the instructor, and using readily available document resources, the course manager and the ACS Branch Chief wrote the necessary additions in four days.

Other Changes in the Course Material

Other significant changes in the course material were deletions, and they reflect the importance of feedback from the target audience. This book names two types of formative evaluation: the walkthrough, and the pilot. If all the reviews of this Collection effort's earlier design and development stages could be considered walkthroughs, this course's materials certainly did not lack critique from experienced mid-level managers, their superiors—including highest level executives. Those reviewers, and the experienced task force members, wanted to ensure that all the competencies the participants needed were covered.

The feedback that the actual target audience—in this case, the participants of the pilot—was able to more sharply define was what the target audience is able to do without instruction. Introductory material the task force and ERC members had considered important was already familiar to the target audience and was therefore eliminated. Material was shortened and rearranged substantially for two lessons—Equal Employment Opportunity and Problem Resolution Program.

Instructor Evaluation

For the pilot, each instructor received a written performance evaluation. The lead instructor evaluated the three other team members on instructional skills, technical competence, utilization of time and administrative ability. The lead instructor was evaluated by the Western Region project manager on the same categories. The executive instructor did not receive a written evaluation.

The Evaluation Phase

Following the pilot, regular conducts of the functional mid-level management training for Collection received summative, but not followup, evaluation. There were evaluations of both the course design and instruction. Learner mastery was also assessed but not formally recorded.

Course Design and Instruction Evaluation

At the completion of each conduct, trainees filled out evaluation forms on which they rated on a scale of five how beneficial the course had been. They also responded to questions regarding the course's design, most beneficial lessons, additions and deletions, group activities, instructors and site location.

To date, there has been one full conduct of this course since the pilot. Because there had not been a functional "how-to-manage-work" course before, the majority of participants have been in their mid-level positions for longer than the designated target audience (i.e., newly appointed mid-level managers who have been in their positions for less than a year). Nevertheless, the course received a 4.32 from the first conduct's 19 participants.

The National Office course manager does a written evaluation of each conduct and its instructors. The course manager visits each conduct at least once, meeting with the participants as a group without the instructors and meeting with the instructors individually. Instructors supply the course manager with all extra materials they used in the conduct. Extra materials may include refinements of existing materials; frequently, they include new material necessitated by changing laws. The extra materials are not only passed on to the instructors of the next conduct, but also are used in the formal revision in 2-3 years.

Learner Mastery Evaluation

This course had an overall assessment of each student's performance which is not routinely documented. As mentioned earlier, the instructors read the participants' written reviews of a subordinate manager. This activity is a culmination of previous lessons.

The evaluation of the Collection Mid-level Functional Management Training Program, on an ongoing basis, focuses on assessing the effectiveness of the course design and the quality of instruction. Such evaluations are traditional within IRS Training and Development. Learner mastery is more informally assessed in this management training program. It should be noted, though, that for the majority of IRS training, especially recruit training, formal testing and evaluation of an individual's performance is the rule, not the exception. Extensive test banks are developed, kept secure, and appropriately updated for technical/professional training programs. Care is given to tie test questions to instructional objectives and training. The magnitude of this care and the extensive efforts expended in course development and revision can be appreciated when the frequency of change of the content for many courses—tax law—is considered.

Summary

This summary highlights aspects of IRS Training, in general, and this case study, in particular.

Use of the Project Agreement

It should be emphasized that IRS Training and Development traditionally uses all five phases of the ISD model put forth in this book in its course development efforts. As stated at the beginning of this chapter, program development at IRS is a study in coordination of communication and resources.

The importance and effectiveness of project agreements cannot be overemphasized. Project agreements force a training program manager to think through the means for accomplishing the five program development phases. This requires delineating functional client involvement and securing the function's agreement to that involvement before a project begins. The delineation of roles and responsibilities not only helps assure different types of functional expertise at critical points within the development process (e.g., task force participation and functional review of draft materials), but also sets forth the professional role and responsibilities of training personnel.

Two organizational realities are central to the effectiveness of a training program's project agreement: Training is a separate division within IRS, having a staff dedicated to the development and delivery of courses; and Training has its own budget (e.g., travel and per diem) critical to both the development and use of programs. This specialized identity promotes professionalism within an organization's training activity, which is very important in an organization such as IRS.

The functional organization components of IRS value training and substantively support training activities with technical expertise during both training development and delivery. The functions also respect Training's efforts to evaluate functional personnel on criteria for training competencies; disruptive or ineffective task force members or technical instructors are dismissed or are not invited back.

Steps Omitted in the Micro Course-Level Development

There are only two steps within the overall training development process that were missing during the development of the Collection course: (1) some form of early formative evaluation (e.g., a walkthrough) with members of the target audience before program pilot, and (2) the incorporation of some form of follow-up evaluation.

Many programs at IRS Training do undergo "developmental testing" with a handful of representative trainees before a pilot class is held. In the case of the Collection functional management training course, this step was short circuited because of the abbreviated timeframe mandated by a December pilot. The inclusion of unnecessary introductory material during the pilot was a result of this omission. Nevertheless, because the pilot class was used as a revision tool, and not as a sanctioning mechanism, feedback from the participants and instructors was used to do the all-important formative fine-tuning of the course.

A few courses (e.g., the "how to manage people" mid-level management training course) currently have follow-up evaluation as an ongoing part of the overall management. Such evaluations of management training courses have been an exception rather than a rule. In recent years, the advent of computer technology has made the possibilities of training and career development follow-up more feasible.

The IRS has been actively exploring computer technologies and acquiring systems (e.g., the Automated Training System, which includes both computer-based training and extensive computer management of all IRS training). Such systems will make follow-up evaluations more of a reality. They will enable the IRS to track an individual's career path, assessing the effectiveness of the training that the organization has provided.

The Involvement of Technical/Professional Personnel in Training

Although Training and Development has been a separate division within the IRS, the training personnel are usually not involved as instructors for most courses. Rather, the training staff specializes in managing the training development and revision process, the logistics for conducting training and the evaluation of courses.

Courses are written and delivered by technical practitioners. The IRS makes extensive use of task forces in the development of training materials. Care is usually given to have nation-wide representation in these groups. Such composition promotes the development of truly representative materials for use service-wide and helps assure acceptance of the materials.

I believe that for technical fields it is always more efficient to have subject-matter experts draft training materials. It is ludicrous to presume that an instructional designer can produce materials that will convey the subtleties of language and relationships of specialized fields. Professionals work long and hard to gain the discriminating mindset of their field. This should be capitalized on, not diluted.

Instructor Training and Evaluation

Even though technical, and not training, personnel deliver most training at IRS, the training division is far from uninvolved. Training trains the instructors (i.e., Basic Instructor Training) and provides the opportunity, coaching and resources for instructors to prepare for each teaching assignment. Training also evaluates the effectiveness of instructors each time they teach a course.

The IRS also has a number of courses that uses outside contractors as instructors (e.g., the "people" course for mid-level managers). Training secures these contractors through appropriate procurement procedures and evaluates their effectiveness no less rigorously than it evaluates IRS instructors.

In many organizations, there are permanent technical personnel on the training staff who conduct, and even develop, training. In contrast, the IRS uses current practitioners. This, I believe, helps ensure up-to-date and vital training. This use of current practitioners also helps maintain the professionalism of the training division; the training staff is not used as a holding station for certain technical personnel. On the contrary, the few technical personnel who have 2-3 year assignments in training are there through competitive selection and for career development.

The Mid-level Functional Management Training Program for Collection

The mid-level functional management training course for Collection was in the warehouse, ready for service-wide use, in the spring of 1985. This was almost seven years after the then newly created Management Training Advisory Committee recommended that the Management Program Section conduct a needs assessment for all three levels of management training.

The fruition of the charge and its subsequent recommendations, which were extensive, are a testament to the effectiveness of the project agreement mechanism and the importance of appropriate involvement of all affected groups within the organization. Such agreement and involvement sustained the effort over years and through many personnel changes.

In the case of the Collection mid-level functional management training program, continuous executive review proved very beneficial. The constant involvement of the Management Training Advisory Committee made possible this program's deviation from the prescribed 1980 Redesign specification for functional training. The Collection course design changed the on-the-job training and seminar classroom to a more limited pre-class memorandum and more extensive "mega-lesson" seminar. At the same time that it allowed a particular deviation from the prescribed approach for functional training, MTAC continued its support of the overall redesign of mid-level management training.

The involvement of the specially formed Executive Review Committee for the Collection effort has been crucial to the success of the program. Suggestions from the special perspective of broad-based executives had a great impact on certain topics. Even more than specific content input, their personal investment has been pivotal to the program's success. For instance, the Assistant Commissioner for Collection was the chairman of the ERC, which was officially constituted only for the duration of the development effort. However, because of the difficulty in securing instructors with Automated Collection System experience, the Assistant Commissioner personally became involved in requesting these instructors from the field. Another example: the two functional members of the ERC attended the last two debriefing days of the pilot. Such involvement had been allowed, but not specified by the Deputy Commissioner in the formation of the ERC.

The message of extensive executive involvement was not wasted on either the Collection function, in general, or its mid-level manager in particular. This course was important enough for the Deputy to appoint executives, and the executives cared enough to truly extend themselves.

Final Thoughts

Too often, training courses are developed by instructional designers in isolation or, at best, with limited client contact. Clients, for the sake of speed and convenience, often would like to "Let Training do it . . . I will tell Training what I want, and they will produce it." Instructional designers can also believe that "It is easier and quicker to work by myself."

Such an approach usually results in a product in a relatively short time. However, speed and efficiency have, more than likely, been bought at the price of richness of product; and, more importantly, acceptance and use of the product. It is human nature to want to use those things in which you have personal involvement. Personal involvement is easier to accomplish when individual players are indeed individuals and, consequently, can be readily identified. Large organizations are harder to involve. Nevertheless, true client ownership can be attained by systematically involving all appropriate hierarchies and levels. To do this, agreement of such involvement needs to be formalized early in the life of the project.

Successful program development cannot occur overnight, and training is no exception.

Credits

The author is grateful to Cheryl Domecq de Gomez, Eleanor Fischer-Quigley, John Moore, Mary Ann Ruth, Emmy van Stolk and the late Linda Harris who either through personal communications or their program products helped me construct the macro program history; and to Barbara Swanson and Clyde Morse who coached me in IRS Training and Development procedures. I would like to express great appreciation for Jack Porter who as program manager brought the micro program to completion and generously conveyed that history.

References

Analysis of mid-level management training needs. Program proposal presented to the Management Training Advisory Committee, Internal Revenue Service, Washington, DC, August 1980.

Training and development (1982). *Internal Revenue Manual. MT 0410.* Washington, DC: Internal Revenue Service.

van Stolk, E.J. (1983). Questionnaire and task force: a comparison of two methods of task analysis (Doctoral dissertation, University of Cincinnati, 1983).

Appendix C

Instructional Systems Development in a Large Computer Company

Richard Duggins

Introduction

In the commercial/industrial environment, many factors influence and modify the implementation of ISD theory. It is well to examine some of these, for tradeoffs and judgmental calls are a fact of life. To anticipate them is to enable the most practical application of theory.

To keep the discussion within reasonable bounds, we will choose as an example a large corporation engaged in the business of information processing. We will further limit the discussion by dealing only with creation of the skills necessary to service the company's products, which have been sold and installed in various widely scattered locations. You may feel that to eliminate consideration of training of development engineers and of salesfolk and others unwisely restricts the discussion; you'll soon see that little of the practice of course development theory is omitted.

Motivation

There is a motivational factor at the ultimate root of decisions in the corporate environment: one that is shared with the public and the military environments. This motivation is more directly brought to bear in the commercial environment than in the other two. Let's examine it.

There is a word used by physicists to describe one action in a system. It's "entropy." Webster's New Ninth Collegiate Dictionary defines it as, among other things, "the steady degradation or disorganization of a system or society." They held that any system, be it an individual, a company or the solar system, must have an input of energy if it is to survive and to continue to operate. In order to prevent the quiet dissolution of the system, there must be a function which monitors the system's performance and determines where an application of energy is required. Our company, as a system, needs to have a feedback mechanism and the capability to apply energy where feedback dictates it. The company must conduct needs analysis.

Linking Training to Organizational Needs

Macro Analysis

So, we've avoided that emotional term "profit motivation," and instead seen that there must be needs analysis for a company to survive. The success of needs analysis and subsequent actions may be measured more closely in the commercial environment than in either the military or the public sector. Certainly it can be measured more quickly and with more agreement, for one need only look at the company's annual report. And we've underscored the importance of needs analysis: it's a matter of survival.

Needs analysis occurs at many levels in the company, starting at the very top. At that level, three actions usually occur. First, there is a careful analysis of existing customers. By surveys and by visits, the company will determine how its products are being used, what the customer thinks of the product's versatility and reliability, and what needs the customer feels are not being met. Second, a competitive analysis group will pore over the competitor's product. They will be interested in its capability and reliability, as well as its price. They may buy and install the competitor's product and subject it to rigorous testing, and also evaluate the technology used. And, thirdly, management will look to their own research and development group. This third step is to see what is possible with the technology as it is understood and practiced in their own shop. If there is a way known to the company's people by which a given function can be performed better, more quickly, or at less cost than the competitor's product, top management should know. Perhaps more important is the opportunity, through new knowledge, to create a new market: to provide a function which the customer will buy and for which there is initially no competition. From this rigorous (and expensive) gathering of information, a decision may come to design, build, and market a new product.

Please understand that this is a very dynamic process. The competitors are going through the same routine; the research and development folk are pushing boundaries back even as the Chief Executive Officer ponders what action to take. As an example of volatility there was a case where a particular company introduced and marketed very successfully a certain computer. As the customers used it, they said "I wish it would do this—or that." Their comments found their way back to the plant of manufacture, and were duly noted. The engineers themselves had ideas about how the product could be improved, and began to design a new level of the product, incorporating all these ideas. Soon they realized that they were not designing a new release level, but were designing a new system! Success is heady, so they pushed on, only to realize that they were infringing on a product area assigned to another group, who were also developing a new system. Now there was internal competition, and many meetings to resolve the dispute. Final resolution came easily, when a directive from the CEO stated that the market trend was to a completely new technology, and that no new product was to use the old! *Both* designs were scrapped.

But let us assume that, amidst the swirling dynamics of change, the decision to design a new product survives. What of needs assessment at this level?

The manager of product engineering to whom the new product is assigned has a host of questions to answer. Exactly what functions are to be implemented? What available technology and technique will best support their implementation? How reliable will the product be? What will it cost to manufacture the product? To sell it? To service it? Will the total cost be within the target price range? What can we do to get the cost down, so that the product may be sold at a more competitive price, or more profitably? To gather answers, an iterative cost estimating process is used. An initial set of design specifications is circulated. Each affected function estimates their cost, and the costs from all functions are assembled. Now we reach the tradeoff point. "If you omit this feature, we can build the product for this much less." "If you design the product with self-diagnosis capability, then the cost of service will be reduced this much." From this, decisions are made and a new set of product specifications are drawn. Again the cost estimate procedure is gone through, and a product cost arrived at.

As this cycle takes place, each affected function is also making its plans. "If this product is really designed and built (not all are!), what will we need to do?" There's that needs assessment thing again.

Course Needs Assessment

Micro Analysis

A training cost estimate is an integral part of a product cost estimate. Why? Reliability of a product is a key issue. To send a trained person to deal with a malfunction is a costly process, both in terms of personnel costs and of impact on the customer. So the product engineering manager will ask for a service cost estimate, which will include the training cost estimate.

Those who are responsible for the service cost estimate must look at the design specifications for the new product, attempt to judge what types of malfunction might occur (Did a part break? Is there an error in the code? Did the customer attempt an improper useage?) and how to deal with each category of failure. The result is a "service strategy."

There are many variations of service strategy. At one extreme, the plan might call for an on-site person who is trained to deal with any eventuality. Or perhaps the customer will be expected to analyze his problem down to a falling "box," replace that box with a spare, and mail the failing one to a central repair station. The other extreme might call for the customer's system to be linked electronically to a "supersystem," which will receive error indications from the customer system, analyze the indications to

ISD in a Large Computer Company 293

isolate the cause of failure, and transmit a message giving the proper corrective action—all without the customer realizing that there was a problem with his machine. In the middle of the spectrum, the service strategy might call for many people to be trained to handle the most common failures, and a very few people trained to handle the unusual ones.

It is at this point that the training organization usually enters the picture, and that the training needs assessment process begins. What training will be needed to implement the service strategy?

The training needs assessment can often be quite crisp because of the work which has preceded it. To establish the training objectives may require only that the trainer understand the product specifications and the service strategy. But two things must be recalled: first, that the whole process is very volatile and subject to change without notice, and second, that the trainer examining the specifications very likely has had no training in the ISD process.

It is an intuitive belief held by many that the best person to train for a new skill is one who has performed that skill, or a similar one, successfully. For example, we believe that a successful salesperson will be able to train other salespersons to be equally effective. The fact that the skills of the successful salesperson have little to do with training skills is not considered. For someone to carefully examine the qualities and techniques of that succssful salesperson, and then to ask "What change must my trainees exhibit if they are to resemble this model?" is not the rule.

Further, the trainer so selected will most likely view his assignment as a career step, not a career. There will always be a few notable exceptions; people who really enjoy their assignment, recognize their training deficiencies, and set about to correct them. Too, the company may recognize the deficiencies. Many have, and some have launched extensive programs to convert successful technicians into instructors and course development people. But to hire a cadre of trainers is not usually an acceptable alternative. To do so would be to create an "overhead" group which would be inflexible in times of workload fluctuation. And so we will continue to have "trainers" who are chiefly interested in preparing for their next career step.

There is an unfortunate, but not uncommon, result of using product technicians for trainers. That is the definition of "exit skills," which aren't really skills at all, and which may stand in isolation from the main body of the material. Watch for skill definitions which begin with "know" or "understand." They are there because the technician is able to deal with a particular problem, and sees that to "know, given a memory printout, where to find a certain string of bytes" is essential to the process. But the *trainee* does not see that! If the knowledge is presented, like an iceberg, in splendid isolation, the student may know it for a time. And, since it is isolated from other association, it, like the iceberg, will with time melt away. Rarely does a company pay people to know, or to understand, or to appreciate things—the company pays people to *do* things!

Another effect of poorly defined task analysis shows up when the developer attempts to develop training objectives from them. It is frequently found that the "exit level skill" is actually an enabling level for a true exit level. This may lead to a political confrontation with the service planner, who wants *his* words in the definition of the course outcome. The ISD person must be a diplomat. Still, regardless of ISD skill, the trainer must consider the product specifications and the service strategy and decide what skills the graduate of the proposed training program must possess.

Training Objectives

With a task analysis completed and exit-level skills defined, attention turns to the trainee. It is well to recall that product development is usually evolutionary: by small increments rather than by huge leaps. It follows that the trainee likely possesses a large part of the skills identified in the task analysis when he begins training. Certainly, when we break the terminal skills into enabling ones, we soon find skills that are already in the trainee's bag of tricks. It's both expensive and demoralizing to repeat that part, so the training objective must include only what's really needed.

ISD in a Large Computer Company 295

But the audience definition can present a trap! Each product manager will specify that the people to be trained for his product be those whose prior training most closely resembles his requirements. There are, however, two problems. First is that there are only a finite number of these people, and they can be spread only so thinly. Depletion of the skills pool may considerably alter the profile of the audience, and the trainer should know to watch for this change. Secondly, these people were performing productively before they were selected for this new training. That work which they were doing may not go away, and it may be necessary to train more people as backfill for them.

So skills pool management becomes an important consideration, and may change the training audience profile, or even the service strategy, and so the training objectives, considerably.

Training Strategy

With realistic training objectives established, how shall we achieve them? It is here that a lack of ISD skills frequently proves to be costly, for the range of choices is wide. In truth, the matter of training strategy seems to be more often an assumption or a default than it is a reasoned decision. Consider only the mode of delivery. For some objectives, the reading of a single printed page may be sufficient. Other objectives may require the development of manipulative skills. Here we must decide whether to bring the student to the equipment or the equipment to the student. Still another case may require the delivery of much new information, with provision for appropriate practice. Shall we bring the student to a central location and have an instructor deliver lecture-lab training? Perhaps computer-based training, allowing the student to stay home, will be sufficient.

Personal observation says that one of two things will occur. The first is that the developer will choose the delivery mode with which he is most comfortable. The second is that the developer will succumb to the glamor of technology and build full-color computer controlled video. The requirements of the objectives fall by the wayside in either case. The example where conscious attention is paid to development of a training strategy is fairly unusual.

Even more unusual is the strategy developed from and supported by a "knowledge tree."

Recent years have introduced another little problem for training strategies. It's learning retention. Technology has reduced failure rates to the point that the trained person may not apply the new skill for months or years. When it's needed, the new skill may have atrophied to the point of uselessness. The training strategy must provide for refresher training. Better yet, the planners may revert to a higher level in the planning and eliminate the requirement for the training by changing the product design or the service strategy. This is an excellent point for discussion of tradeoff of training cost against further engineering cost.

Course Development

The most common difficulty found in course development is for a "developer" to start writing a course with little or no attention to the preceding steps. You'll recognize the problem when you see the developer take a yellow scratch pad, write the name of the course at the top, and then start writing text. This is not as silly as it seems when you consider that ISD is really an infant of a discipline—not yet recognized by much of the training community. Add to that the fact that the rate of change of skills is accelerating furiously, involving many people in the need for training. Those trained in the ISD discipline must be true salespersons to show the need for organized ISD.

With a proper strategy in place, it is time to develop course material. This requires knowledge of the subject matter, skill in selecting and using the best media for the topic at hand, excellent communication skills (including at least adequate command of the English language!), and the ability to work effectively under the time pressure of a product development schedule. There may be a shortage of people with such skills, so not infrequently the prime qualification of those assigned is that they are *available*. One approach is to find several subject matter experts and divide the project among them. Another is to find SME's, instructional designers, and media experts and use a team. Either approach requires a new skill: project coordination. Those two words label a very complex skill. The Project Coordinator must assure communications, keep things on schedule, soothe bruised egos, insulate the team from the demands of the rest of the world, and act as a lightning rod for the bolts from higher management. Such a skill is to be treasured.

There was a personal experience which contributed to my grey hairs, and which can illustrate this point. The project was a significant revision of an existing course. The development site was not at the education location, so the communication problem of distance was introduced. The project team included people from both school and development locations, so to have a team meeting was difficult. One team member was an ardent "type A" individual who was not going to take any guff from anyone, in addition to being eager for promotion. Members of the team from the development location were eager to implement new techniques. And the project coordinator was a seasoned course development person, not trained in project coordination (*mea culpa!*), who tried to reach group consensus on every point. The project completion was months late.

The use of proven technicians as course developers introduces another problem: effective communications within the course material. A survey showed that our audience, in general, read at about the 12th grade level, but that the range was very wide. To increase the reading level of the material one grade level could cause trouble for more than 10 percent of the audience. To this, add the fact that many of the technicians themselves were not really excellent in their command of the English language (in the written form), and you've got trouble. The trouble pretty well demands the attention of a qualified editor who can handle the misspelling, the misplaced commas and run-on sentences, but also an editor who can help make the content palatable. That's a ticklish assignment. The author is proud of that priceless prose. To criticize is to call someone's baby ugly.

Effective written communication also depends on the worth of earlier efforts to analyze terminal skills, and to organize the resulting enabling skills into a valid learning sequence. If the trainee is being lead carefully from the known to the new; if he is adding new knowledge, much in the way of poor sentence construction may be excused. But if facts are being presented in a random fashion using muddy, redundant and ambivalent wording, learning will be hindered.

The person(s) who are assigned to develop the technical content of the course have a peculiar handicap; the product for which they are to develop training does not exist! It's in development, and like a fetus in the womb, it changes daily. There is no one to train course developers on the product so that they might write the necessary material—no one knows the product yet. Many people know a great deal about small portions of the project, because product development also follows a team approach. The course developer must seek out fragments of knowledge here and there and piece them together to create a total understanding. In fact, it is common for the course developer to know more about the total product than the engineers who designed and built it!

There's another facet to this hazard. There is a product development deadline; the company's annual and long-range plans depend on having a product to sell by a given point in time. It follows that the development engineers feel pressure to meet deadlines. As a result, they may view the trainer who is seeking knowledge as an unwelcome interruption, and give him as little attention as is possible.

This "do it yourself" training of the trainers presents a special hazard in the training sequence which is to use complex media. The production of interactive video demands a broad spectrum of specialized skills; skills which are not common in the job market. Also, the mechanics of production are very time consuming. The trainer is frequently faced with the requirement for detailed product knowledge when the trainer is still working at the training strategy level. So the pragmatic trainer will do two things—ring up the media folk very early in the course development cycle, and not plan for complex media in the first few classes.

Course Validation

Course validation is one part of the ISD process which seems to be instinctively accepted, although it may be short-changed in the press of business. Techniques commonly used are peer reviews (sometimes called walkthroughs) as the material is being developed and small test classes with those students who will service the first products delivered. Subject matter experts will be used for review also, when the engineer on a particular facet of the product can find time in his schedule to do the review. End of the Course surveys are very commonly used, but they tend to have a strong "halo" effect. The student sees the end, and he wants to be through! Less frequent, though not uncommon, are follow-up surveys sent to the ex-student, who has returned to his regular assignment and who has had opportunity to exercise his new skills. Still less frequent is the visit by the trainer to the ex-student at his regular workplace, where he can both discuss and observe the use of the new skills.

The subject of course validation cannot be left without some discussion of "tests." Test marks in the 90s on paper-and-pencil tests are taken as evidence that the student has acquired the desired behavior patterns. There are two large problems with this. First, the student is not, in most cases, being trained to take tests, and any judgment of his skill level based on test results can be only inferential, not directly measured. The second problem is that there isa regrettable tendency to forget course objectives when the test is being prepared. That which is tested may or may not be germane to the objectives.

Course Implementation

There is one aspect of course implementation which will be singled out for discussion: the need for on-going monitoring of course results. Ths is necessary because of the climate of change. That which is most likely to change is the audience profile. First classes will normally have battle-hardened veterans, while later ones will have less experienced people. One way to monitor this is the End of the Course survey, administered as the student walks out the door for the last time. One should realize, however, that other influences ("I want to go home!") may influence the response as much as the appropriateness of the training. Another way to monitor the training is the work location survey mentioned earlier. This tends to be the more reliable of the two. Still another way is to monitor actual product performance, and to infer that good product performance is the result of good training. Rest assured that others will question the training if the product performs poorly! But to infer that good product performance is the result of good training is poor business. It is quite possible that the product requires no service, and therefore any training is a waste of money!

Course Revision

There is a difference between "revision" and "correction." Correction is the action taken to correct typos or the unintentional omission of page 13 or the ambivalent wording in another place. Revision, on the other hand, means that a significant rework of the course is to be done.

A first reason for revision is the proof that an initial assumption was wrong. By way of example, there was the machine which used magnetic drum storage. The read-head-to-drum clearance was quite critical, and the initial assumption was that, after factory adjustment, readjustment was not necessary. Therefore, students were not trained to make the adjustment. A very short time in the field proved that assumption totally wrong; readjustment *was* necessary. So there was a hectic period where factory people made field trips to make the adjustments, and a frantic revision of the course was ordered to develop the skill to adjust.

A second reason for course revision lies in the fact that product development does not stop when the first machine is shipped. Better ways may be found to perform existing functions, or new functions may be added. New service aids may be provided, or new procedures developed. There is a need to determine whether these changes can be handled by revision or whether a new training course is required. There is also the need to examine the expected life of the product and the resulting training load, to determine whether revision can be justified financially.

Full Circle!

The word "need" was used in the last couple of sentences to make a point. We've gone full circle in the training process, and are trying to do needs assessment for training requirements. The product development manager is doing needs assessment for his product. The CEO is listening to the customer and to the market analysis folk and to the R&D people, trying to assess the needs of the business, in order that it may survive. The process is ongoing, and the need for a systematic approach is always with us.

References

Andrews, D.H., and Goodson, L.A. (1980). A comparative analysis of models of instructional design. *Journal of Instructional Design, 3*(4), 2-16.

Angrosino, Michael (1976). The Indian cinema in the West Indies. *The Third World Review, 29*(1), 75-78.

Ausubel, D.P., Novack, J.D., and Hanesian, H. (1978). *Educational psychology: A cognitive view* (2nd ed.). New York: Holt, Rinehart and Winston.

Beal, G.H., and Rogers, E.M. (1960). *The adoption of two farm practices in a central Iowa community.* Special Report 26. Iowa Agricultural and Home Economics Experimental Station.

Beal, G.H., Rogers, E.M., and Bohlen, J.M. (1957). Validity of the concept of stages in the adoption process. *Rural Sociology, 23*(2), 166-168.

Berman, P., and McLaughlin, M.W. (1978). *Federal programs supporting educational change, vol. VIII: Implementing and sustaining innovations.* Santa Monica, CA: Rand Corporation.

Branson, R.K., Rayner, G.T., Cox, J.L., Furman, J.P., King, F.J., and Hannum, W.H. (1975). *Interservice procedures for instructional systems development.* Ft. Monroe, VA: U.S. Army Training and Doctrine Command.

Briggs, L.J. (1982). Systems design in instruction. In H. Mitzel (Ed.), *Encyclopedia of educational research* (5th ed.). Washington, DC: American Educational Research Association.

Bruner, Jerome (1966). *Towards a theory of instruction.* Cambridge: Harvard University Press.

Campbell, D.T., and Stanley, J.C. (1966). *Experimental and quasi-experimental designs for research.* Chicago: Rand McNally.

Clark, R.E. (1982). Antagonism between achievement and enjoyment in ATI studies. *Educational Psychologist, 17*(2), 92-101.

Clark, R.E. (1983). Reconsidering research on learning from media. *Review of Educational Research, 53*(4), 445-459.

Coffing, R.T., and Hutchinson, T.E. (1974). Needs analysis methodology: A prescriptive set of rules and procedures for identifying defining and measuring needs, *RIE,* ED095654.

Colby, B.N. (1975). Culture grammars. *Science, 187*(4180), 919.

Cummings, O.W., and Bramlett, M.H. (1984). Needs assessment: A maximizing strategy that works for information development. Paper presented at the annual meeting of the Evaluation Network and Evaluation Society, San Francisco, October.

Delbecq, A.L., Van de Ven, A.H., and Gustafson, D.H. (1975). *Group techniques for program planning: A guide to nominal group and Delphi processes.* Glenview, IL: Scott-Foresman.

Deterline, W.A. (1974). *Performance analysis.* Palo Alto, CA: Deterline Associates.

Dick, W., and Carey, L. (1985). *The systematic design of instruction* (2ed). Glenview, IL: Scott-Foresman.

Donaldson, L. & Scannell, E.E. (1985). *Human resource development.* Reading, MA: Addison-Wesley.

Duchastel, P.C., and Merrill, P.F., (1973). The effects of behavioral objectives on learning: A review of empirical studies. *Review of Educational Research, 43,* 53-70.

Fessler, T. (1980). Moving from needs assessment to implementation: Strategies for planning and staff development. *Educational Technology, 20*(6), 31-35.

Fleming, M.L., and Levie, W.H. (1978). *Instructional message design.* Englewood Cliffs, NJ: Educational Technology Publications.

Gagne, R.M., and Briggs, L.J. (1979). *Principles of instructional design.* (2nd ed.). New York: Holt, Rinehart and Winston.

Gagne, R.M., Briggs, L.J., and Wager, W.W. (1988). *Principles of instructional design.* (3rd ed.). New York: Holt, Rinehart and Winston.

Gagne, R.M. (1977). *The conditions of learning* (3rd ed.). New York: Holt, Rinehart and Winston.

Gagne, R.M. (1985). *The conditions of learning* (4th ed.). New York: Holt, Rinehart and Winston.

Hall, Edward and Reed, Mildred (1980). The sounds of silence.. *Anthropology.* Gilrod, CT: Dushkin Publishing Company.

Hannum, W.H. and Briggs, L.J. (1982). How does instructional systems design differ from traditional instruction. *Educational Technology, 22*(1), 9-14.

Hannum, W.H. (1984). Implementing instructional development models. *Performance and Instruction Journal, 22*(8), 16-20.

Hannum, W.H. (1988). Designing courseware to fit subject matter structure. In D.H. Jonassen (Ed.). *Instructional designs for microcomputer courseware.* Hillsdale, NJ: Lawrence Erlbaum Associates.

Hansen, C.D. (1985). A study in cognition and culture through subject-designed stories in film. Unpublished doctoral dissertation, University of North Carolina, Chapel Hill, NC.

Jamison, D., Suppes, P. and Wells, S. (1974). The effectiveness of alternative instructional media: A survey. *Review of Educational Research, 44*(1), 1-68.

Jonassen, D.H., and Hannum, W.H. (1986). Analysis of task analysis procedures. *Journal of Instructional Development, 9*(2), 2-12.

Jones, E.E. and Clay, M.C. (1984). Needs assessment. Paper presented at the annual meeting of the American Society for Training and Development, Dallas, Texas, May.

Katz, E. and Wedell, E.G. (1978). *Broadcasting in the third world.* London: Macmillan.

Kaufman, Roger (1976). *Needs assessment: What it is and how to do it.* San Diego, CA: University Consortium on Instructional Technology.

Kaufman, Roger (1985). Linking training to organizational impact. *Journal of Instructional Development, 8*(2), 23-29.

Kaufman, Roger (1986). Obtaining functional results: Relating needs assessment, needs analysis and objectives. *Educational Technology,* 24-27.

Kaufman, Roger and English, F.W. (1979). *Needs assessment: Concept and application.* Englewood Cliffs, NJ: Educational Technology Publications.

Kemp, J.E., and Dalton, D.K. (1985). *Planning and producing audiovisual materials* (5th ed.). New York: Harper and Row.

Kirkpatrick, D.L. (1987). Evaluation. In R.L. Craig (Ed.) *Training and development handbook* (3rd ed.). New York: McGraw Hill.

Kulik, J., Kulik, C., and Cohen, P. (1980). Effectiveness of computer-based college teaching: A meta-analysis of findings.

Review of Educational Research, 50, 525-544.

Lane, K.R., Craften, C., and Hall, G.J. (1983). Assessing needs for school district allocation of federal funds. Paper presented at the annual meeting of the American Educational Research Association, Montreal, Canada, April.

Linstrone, H.A., and Turoff, M. (1975). *The Delphi method: Techniques and applications.* Reading, MA: Addison-Wesley.

Maanen, J.V. (1979). The process of program evaluation. *The grantsmanship center news,* January/February, 29-64.

Mager, R.F., and Pipe, P. (1984). *Analyzing performance problems or you really oughta wanna* (2nd ed.). Belmont, CA: Pitman Learning, Inc.

Mager, Robert (1972). *Goal analysis.* Belmont, CA: Fearon.

Mager, Robert (1977). The winds of change. *Training and Development Journal,* October, 12-20.

Markle, David G. (1977). First aid training. In L.J. Briggs (Ed.) *Instructional design: Principles and applications.* Englewood Cliffs, NJ: Educational Technology Publications.

Martin, B.L., and Briggs, L.J. (1986). *The affective and cognitive domains: Integration for research and instruction.* Englewood Cliffs, NJ: Educational Technology Publications.

Mayer, R.E. (1979). Can advance organizers influence meaningful learning? *Review of educational research, 49,* 371-383.

McCormick, E.J. (1979). *Job analysis: Methods and applications.* New York: American Management Association.

McKeachie, W.J. (1974). The decline and fall of laws of learning. *Educational Researcher,* March, 7-11.

Merton, R.F. (1978). Resolution of conflicting claims concerning the effect of behavioral objectives on student learning. *Review of Educational Research, 48*(2), 291-302.

Messick, S. (1984). The nature of cognitive styles: Problems and promise in educational practice. *Educational Psychologist, 19,* 59-75.

Patterson, C.H. (1977). *Foundations for a theory of instruction and educational psychology.* New York: Harper and Row.

Provus, M.M. (1971). *Discrepancy evaluation.* Berkeley, CA: McCutchen.

Reigeluth, C.M., and Stein, F.S. (1983). The elaboration theory of instruction. In C.M. Reigeluth (Ed.), *Instructional design theories and models: An overview of their current status.* Hillsdale, NJ: Erlbaum.

Reigeluth, C.M. (1983). *Instructional design theories and models: An overview of their current status.* Hillsdale, NJ: Erlbaum.

Rice, E.G. (1980). On cultural schemata. *American Ethnologist,* 7 (1), 152-153.

Rogers, E.M. (1973). *Communication strategies for family planning.* New York: Macmillan Publishing Company.

Rogers, E.M. (1983). *Diffusion of innovation.* New York: Macmillan Publishing Company.

Rogers, E.M. and Shoemaker, F.F. (1971). *Communications of innovations.* New York: Macmillan Company.

Rumelhart, D.E. and Ortony, A. The representation of knowledge in memory. In R.C. Anderson, R.J. Spiro, and W.E. Montague (Eds.), *Schooling and the acquisition of knowledge.* Hillsdale, NJ: Lawrence Erlbaum Associates.

Saloman, Gabriel (1979). *Interaction of media, cognition and learning.* San Francisco: Jossey-Bass.

Schramm, Wilbur (1977). *Big media, little media.* Beverly Hills, CA: Sage Publications.

Schramm, Wilbur and Learner, Daniel (1976). *Communication and change.* Honolulu: East-West Center.

Scriven, M., and Roth, J. (1978). Needs assessment: Concept and practice. In S. Anderson and Coles, C. (Eds.) *New directions for program evaluation.* San Francisco: Jossey-Bass.

Scriven, M. (1978). Needs assessment is harder than you think. *Report on education research,* 10(25).

Shavelson, R.J. (1981). *Statistical reasoning for behavioral sciences.* Boston: Allyn and Bacon.

Slatter, P.E. (1958). Contrasting correlates of group size. *Sociometry,* 21, 129-139.

Smith, M.E. (1980). How big a sample do I need for my evaluation? *NSPI Journal,* 19, 3-8.

Stufflebeam, D.L. (1977). Working paper on needs assessment in evaluation. Paper presented at the annual meeting of the American Educational Research Association, San Francisco, April.

Stufflebeam, D.L., Foley, W.J., Gephart, W.R., Guba, E.G., Hammond, R.L., Merriman, H.O., and Provus, M.M. (1971). *Educational evaluation and decision making.* Itasca, IL: Peacock.

Tracey, W.R., Flynn, E.B. and Legree, C.L. (1970). *The development of instructional systems: Procedures manual.* Fort Devans, MA: United States Army Security Agency.

Tyler, R. (1969). *Educational evaluation: New roles, new means.* Chicago: National Society for the Study of Education.

Weaver, R.L. (1981). The small group in large class. *Educational Forum, 48,* 65-71.

Witkin, B.R. (1977). Needs assessment kits, models and tools. *Educational Technology, 17*(11), 5-18.

Worth, Sol and Adair, John (1975). *Through Navajo eyes.* Bloomington: Indiana University Press.

Index

Acceptance pretesting, 217
Accreditation model of evaluation, 196-197
    questionable objectivity of, 197
Achievement, 200, 282-283
Adair, John, 177
Adoption rates
    and compatibility of innovation, 180-181
    and complexity of innovation, 181
    differences in, 182-184
    and observability of innovation, 181-182
    and relative advantage of innovation, 179-180
    and trialability of innovation, 181
Adoption styles, 183-184
    early adopters, 183
    innovators, 183
    laggards, 184
    majority adopters, 183
    traits of willing adopters, 184
Advance organizers, 133-134
Agrosino, Michael, 153
Alterable models, 213
Analysis, likely changes in, 216-218
Analysis of records, reports, and work samples, 58
Anderson, C.L., 225
Andrews, D.H., 159, 239

Anthropological methods, 210-211, 212
Attitude domain, 128
    goal analysis for, 128
    sequencing instructional content for, 131
Ausubel, D.P., 10, 116, 133, 134

Back, S.F., 240
Baker, Meryl, 225, 232, 233, 249, 250, 252
Baseline performance data assessment, 48, 55
Beal, G.H., 184
Berman, P., 10
Big Media, Little Media, 140
Bramlett, M.H., 53
Branson, Robert K., 82, 225, 229, 240, 249, 252
Briggs, L.J., 5, 16, 24, 133, 135, 153
Brooke, Martha, 255
Bruner, Jerome, 176

Campbell, D.T., 201
Carey, L., 5, 111, 248
Case studies, 144
Catalysts, training designers as, 174
Change agents, training designers as, 172, 175, 184
Clarification stage of data collection, 54
Clark, R.E., 119, 152

309

Clay, M.C., 39
Coffing, R.T., 41
Cognitive task analysis, changes in, 216
Cohen, P., 154
Colby, B.N., 177
Communication, selectivity in, 177
Communication lines, 174-176
    models for, 175
Communication problems, arising from cultural perspectives, 176-177
Compatibility of innovation, 180-181
Complexity of innovation, 181
Computer-assisted instruction, 149-150
Computer-based control of delivery, 221
Computer-based instruction
    for Job Skills Education Program (JSEP), 226, 234, 242
Computerized media, 149-150
    advantages and disadvantages of, 151-152
    best use of, 150
    costs of, 150
    reducing learning time, 154
Computer-managed instruction, 149
Conditions component of an objective, 107, 108
Configuration management, 239
Content development, 34-35
Content elaborations, providing, 135
Content overview, providing, 133
Content validity, 238
Context evaluation, 195

Continuing professional education, for Internal Revenue Service macro program, 260-262
Cost effectiveness
    contributions of technology to, 234-235
    and methods and media selection, 155, 156
Course-based models, 239
Course design evaluation, for Internal Revenue Service micro program, 282
Course design phase, for Internal Revenue Service micro program, 268-270
Course design review, for Internal Revenue Service micro program, 274-276
Course development, 296-298
Course development phase, for Internal Revenue Service micro program, 276-277
Course implementation, 299-300
Course implementation phase, for Internal Revenue Service micro program, 277-279
Course length, 111
Course revision, 300
Course validation, 298-299
Cox, J.L., 229
Craften, C., 39
Criterion component of an objective, 108-109
Cross-functional executives, in the Internal Revenue Service, 256
Cross-functional training, for Internal Revenue Service macro program, 260, 261
Cultural bonds, 176-178
Cultural factors, influencing training design, 176-178

Index

Cultural grammars, 177-178
Cultural perspective, 210-211
    and communication problems, 176-177
Cummings, O.W., 53
Curriculum standardization, 234, 235

Dalton, D.K., 5
Data analysis, 61-64
    to identify training needs, 61-62
    and performance analysis, 62
    for prioritizing training needs, 63-64
Data collection for evaluation, 201-205
    experimental designs, 203-204
    pre-experimental designs, 201-203
    quasi-experimental designs, 204-205
Data collection for needs assessment
    analysis of records, reports, work samples, 58
    clarification stage, 54
    by consulting with other organizations, 56-57
    data gathering needs, 54-55
    determining sample size for, 59-60
    field procedures, 61
    general guidelines, 51-52
    interviews used for, 56, 57
    key informant stage, 52
    methods and sources for, 56-59
    for needs analysis, 51-61
    observations used for, 56, 58
    questionnaires used for, 58
    specific guidelines, 52-60
    structured groups used for, 53, 57-58
    tests used for, 57, 59
    validation stage, 53-54
Decision making model of evaluation, 195-196
Deductive learning style, 119
Delbecq, A.L., 53
Delphi method, 53
Demonstration, 143-144
    advantages and disadvantages of, 147
    cost a problem of, 144
Design, 33-34
    See also Training design
    of instructional displays, 219
    macro design, 28, 33-34, 96-104
    micro design, 29, 33, 34, 104-109
Design decisions, based on learning research, 220
Deterline, W.A., 69
Developing countries, advantages of instructional systems development for, 6-7
Development, likely changes in, 220-221
Development costs, 20-21
Development phase, of military personnel training programs, 236-239
Dick, W., 5, 111, 148
Discussion, 142-143
    advantages and disadvantages of, 146
    best group size for, 142
    problems with, 142
Domains of learning, 126-129

analysis techniques for, 126-129
attitude domain, 128-129, 131
information domain, 127-128, 129-130
intellectual skills domain, 127, 130
motor skills domain, 127-128, 130-131
sequencing instructional content for, 129-131
Donaldson, L., 153
Duchastel, P.C., 112
Duffy, T.M., 226
Duggins, R., 289

Early adopters, 183
Education and training, need for improved approaches to, 4
Elaboration analysis, 126-127, 128
English, F.W., 39
Entering characteristics, 117-122
assessing, 120-122
general characteristics, 118
learning styles, 118-120
specific entry behaviors, 118
Essential prerequisites, 107
Estimation, 159, 160-165
experience vital for, 160-161
Gantt charts used in, 162, 163
importance of, 160
PERT diagrams used in, 162, 164-165
of time needed, 160, 161-165
Evaluation, 36-37, 191-205
accreditation model, 195, 196-197
of achievement, 200
approaches to, 194-199
context evaluation, 195

data collection for, 201-205. See also Data collection for evaluation
decision making model of, 195-196
distinguished from measurement, 192-193
of education/training outcomes, 199-200
follow-up, 36-37
formative, 36-37, 245, 274-276, 277
goal based model, 195, 197-198
goal free model, 195, 198-199
of impact of training, 200
importance of measurement to, 192-193
influenced by anthropology and sociology, 212
input evaluation, 195-196
for internal Revenue Service micro program, 274-276, 280, 281-283
of job performance, 200
likely changes in, 223-224
of military personnel training programs, 244-252
models of, 192
orientation geared to purpose, 192
process evaluation, 196
product evaluation, 196
purposes of, 191-193
summative, 36-37
of trainees' reaction, 199-200
uses of results of, 194
Examples, importance of, 134
Exercises, importance of providing, 136

Index

Experimental designs, 203-204
Expert-novice distinctions, 217
Expert systems, as job aids, 217, 219
Explicit model of instruction, 242

Farr, Beatrice J., 225, 247, 249, 252
Feasibility analysis, for methods and media evaluation, 140-141
Feedback, 78-79
    importance of providing, 136
Fessler, T., 39
Field procedures for data collection, 61
Fleming, M.L., 5, 134
FLIT program, 226
Flynn, E.B., 64
Formative evaluation, 36-37
    for Internal Revenue Service micro program, 274-276
    for Job Skills Education Program (JSEP), 245
Formative review, for Internal Revenue Service micro program, 227
Frederiksen, J.R., 226
Front-end analysis, 27-28, 30-33
    data collection during, 32. See also Data collection for needs assessment
    for Internal Revenue Service macro program, 258-259
    for Internal Revenue Service micro program, 264-265
    performance analysis during, 32. See also Performance analysis
    task analysis during, 33. See also Task analysis

Functional executives, in the Internal Revenue Service, 356
Functional training, for Internal Revenue Service macro program, 260, 261
Function-wide survey, for Internal Revenue Service micro program, 266-267
Furman, J.P., 229

Gagne, R.M., 5, 90, 126, 127, 132, 133, 135, 152, 153, 220, 240
Gantt charts, 162, 163
General Accounting Office (GAO), auditing military training programs, 233-234
General characteristics of learners, 118
    assessing, 120-121
Goal analysis, 105-106, 128
Goal based evaluation model, 197-198
    narrow focus of, 197
Goal free model of evaluation, 198-199
Goals and objectives, 93-114
    See also Objectives
    as basis for course content and sequence, 106-109
    functions and benefits of, 109-110
    goal analysis to clarify, 105-106
    guidelines for developing, 109-112
    need for clarity in, 105-106
    and participant progress monitoring, 109
    research data on, 111-112

role in content selection and sequencing, 109-110
role in evaluation, 110
selecting, 110-111
Goodson, L.A., 159, 239
Group instruction, 221-222
Guidance, providing, 136
Gustafson, D.H., 53

Hall, Edward, 177
Hall, G.J., 39
Hamovitch, M., 225, 249, 250
Hannum, W.H., 9, 16, 84, 229
Hansen, C.D., 154
Harding, S.R., 225, 232, 238
Huff, K., 225, 233, 249
Hutchinson, T.E., 41

Implementation, 35-36
  See also Implementing planned change
  instructor training during, 36
  likely changes in, 221-223
  need for monitoring, 299-230
  shaped by market research, 211-212
  staged approaches to, 35
Implementation phase of military personnel training programs, 240-244
Implementation plan, 72, 184-189
  advantages of staged approach, 185
  continuing planned considerations, 187-189
  determining adequacy of support linkages for, 188
  maximizing resource use, 188
  piloting the program, 185-187
  reviewing personnel needs, 188
  reviewing success criteria, 189
  reviewing training needs, 187-188
  staged approach recommended, 184-185
Implementing planned change, 171-190
  See also Implementation
  designers as change agents, 172
  developing implementation plan, 184-189. *See also* Implementation plan
  factors influencing target audience, 176-184. *See also* Target audience
  importance of communication lines, 174-176
  questions for consideration, 171-174
Implicit model of instruction, 241-242
In-baskets, 144
  for Internal Revenue Service micro program, 271, 273
Incentives, 78, 79
Independent study, 143
  advantages and disadvantages of, 146
Indicator statements, 229
Inductive learning style, 119
Information domain, 126-127
  elaboration analysis of instructional content for, 126-127
  sequencing instructional content for, 129-130
Innovation
  attractiveness of, 178-182

Index 315

characteristics of influencing adoption rates, 179-182
compatibility of, 180-181
complexity of, 181
observability of, 181-182
relative advantage provided by, 179-180
trialability of, 181
Innovators, 183
Input evaluation, 195-196
Instruction, implicit and explicit models of, 241-242
Instructional analysis, new methods for, 218
Instructional content organizing, 123-129
Instructional content sequencing, 129-131
    for attitude domain, 131
    better models for, 218
    for information domain, 129-130
    for intellectual skills domain, 130
    for motor skills domain, 130-131
Instructional displays, design of, 219
Instructional events
    describe objectives, 132-133
    describe prerequisite knowledge, 135
    elaborate content, 135
    provide examples, 134
    provide exercises, 136
    provide feedback, 136
    provide prompts and guidance, 136
    providing advance organizers, 133-134
    providing content overviews, 133
    sequencing of, 131-137
Instructional requirements, algorithms for identifying, 219
Instructional Systems Development (ISD)
    for a computer company, 289-301
    applications of, 6-7
    as a support function, 169
    as a team effort, 158, 166-170
    benefits of, 15, 18
    compared with traditional instruction, 16-18, 19-20
    costs of, 13
    design levels for, 94-95. *See also* Design; Training design
    development costs for, 20-21
    economic advantages of, 6-7
    emphasis on outcomes important for, 21-22
    encouraging corporate culture, 8
    future developments for, 207-224
    illustrated, 225-252, 255-288
    for Internal Revenue Service, 255-288
        See also Internal Revenue Service
    introduction to, 5-8
    as an investment, 7
    likely changes in procedures, 216-224
    macro design for, 28, 33-34, 96-104
    micro design for, 29, 33, 34, 104-109

model of
See Instructional Systems Development (ISD) model
needing support from other areas in the organization, 21, 169
origins of, 5-6
reasonable expectations from, 11-12
reducing learning time, 6-7
requirements for using, 20-23
research on, 12-18
resistance to, 22-23
role in strategic planning, 7-8
savings resulting from, 13
staff development for, 23
supportive climate for, 21, 169
training program benefits, 15
training time reduced by using, 13, 14, 15, 16
when to use, 23-24
Instructional Systems Development (ISD) model, 25-38
acceptance pretesting for, 217
anthropological methods helping, 210-211
broadening base of, 210-212
cognitive orientation needed, 208
content development phase, 29-30, 34-35
course-based, 239
deficiencies in lesson design, 214-215
design phase, 28-29, 33-34
development of alterable models, 213
distinctions between macro and micro models, 212
evaluation phase, 30, 36-37. See also Evaluation
front-end analysis phase, 27-28, 30-33
future directions for, 208-210
graduated, 215
how-to emphasis needed, 209, 210
illustrated, 225-252, 255-288
implementation phase, 30, 35-36. See also Implementation; Implementing planned change
importance of user acceptance of, 208-209
improvements in, 212-216
instructional strategy improvement needed, 209
iterative vs. linear models, 213
layering in, 214
macro design, 28, 33-34, 96-104
market research shaping, 211-212
micro design, 29, 33, 34, 104-109
outlined, 25-30
procedural flowcharts incorporated in, 216
product driven models, 213
quality control needed, 214
research information shaping, 211
separation of delivery and management, 209-210
system-based, 239
Instructor evaluation, for Internal Revenue Service micro program, 281, 282

Index

Instructor preparation, for Internal Revenue Service micro program, 279-280
Intellectual skills domain, 127
    learning hierarchy analysis of, 127
    sequencing instructional content for, 130
Intended outcomes, describing, 132-133
Intended participants
    See also Target audience
    analysis of, 115-122
    assessing entering characteristics of, 120-122
    assessing general characteristics of, 120-121
    assessing learning styles of, 122
    assessing specific entry behaviors of, 121-122
    entering characteristics of, 117-122
    general characteristics of, 118
    learning styles of, 118-120
    level of beginning instruction for, 116
    needing prior knowledge, 116
    presentation of instruction to, 116-117
    specific entry behaviors of, 118
Interactive media, 221
Internal Revenue Service
    cross-functional executives in, 256
    functional executives in, 256
    importance of project agreements, 256, 257-258, 283-284
    instructor training and evaluation for, 285-286
    involvement of technical/professional personnel in training, 285
    macro program for, 258-263. See also Internal Revenue Service macro program
    management hierarchy in, 256
    micro program for, 263-283. See also Internal Revenue Service micro program
    mid-level functional management program for Collection, 286-287
    redesign of management training for, 255-288
    training team managers for, 256, 257
    training team members, 256-257
    validation of training programs, 259
Internal Revenue Service macro program
    continuing professional education in, 260-262
    cross-functional training in, 260, 261
    front-end analysis for, 258-259
    functional training in, 260, 261
    macro design phase, 259-260
    scheduling, 262-263
    task forces for front-end analysis, 258-259
    training needs identified for, 259-260
Internal Revenue Service micro program, 263-283

course design and instructor evaluation, 282
course design phase, 268-270
course development phase, 276-277
course implementation phase, 277-279
course materials changes, 281
evaluation phase, 282
formative evaluation for, 274-276
formative review for, 277
front-end analysis for, 264-265
function-wide survey for, 266-267
instructor evaluation for, 281
instructor preparation for, 279-280
learner mastery evaluation for, 282-283
major results and revisions in, 280-281
objectives—instructional outcomes for, 270-271
pilot evaluation and revision for, 280
review of course design for, 274-276
service-wide survey for, 267
steps omitted from course development, 284-285
task analysis of Collection management for, 264-265
task forces for task analysis, 264-265, 268, 274
training method selection for, 271-174
Interpersonal skills, teaching, 144-145

Interservice Procedures for Instructional Systems Development Project, 207
Interviews
for data collection, 56, 57
for task analysis, 86
Iterative models, 213

Jamison, D., 152
Job aids, 77
expert systems as, 217, 219
increasing use of, 220
Job analysis, for military personnel system, 229-233
Job breakdown, for task analysis, 84-85
Job design, 78
shaping training design, 212
Job environment problems, 73-74
solutions to, 77-78
Job incumbent surveys, for task analysis, 86
Job Oriented Basic Skills (JOBS) program, 225-252
analysis phase of, 227-233
design phase of, 233-236
development phase of, 236-239
deviations from instructor guides, 244
eligibility for, 228
evaluation phase of, 244-245, 249-251
implementation compared with Job Skills Education Program (JSEP), 242
implementation phase of, 240-244
job analysis for, 232-233
need for, 228

Index

organizational structure of, 237
purpose of, 232, 233
schedule for, 226
skill clusters for, 232
success rate for, 250-251
volunteers for, 228-229
Job redesign, 77
Job Skills Education Program (JSEP), 225-252
  computer-based instruction, 226, 234
  evaluation of, 251
  field testing of, 242-243
  formative evaluation issues for, 245
  implementation compared with Job Oriented Basic Skills (JOBS) program, 242
  instructor and management training for, 243-244
  management of, 240-241
  management plans for, 235
  off-line elements in, 236
  organizational structure of, 237-238
  soldier attitudes to, 247
  task analysis for, 229-232
Job-task analysis
  See also Task analysis and needs assessment, 48
Job task descriptions, 84, 86
Jonassen, D.H., 84
Jones, E.E., 39

Katz, E., 175
Kaufman, Roger, 9, 39, 40, 41, 63
Kemp, J.E., 5
Key informant stage of data collection, 52

King, F.J., 229
Kirkpatrick, D.L., 199
Knowledge engineering methods, 216
Knowledge representation methods, 218
Knowledge/skill analysis, 87, 90
Knowledge/skill problems, 73
  solutions to, 76-77
Kulik, C., 154
Kulik, J., 154

Laggards, 184
Lane, K.R., 39
Learner, Daniel, 176
Learner mastery evaluation, for Internal Revenue Service micro program, 282-283
Learners, general characteristics of, 117
Learning hierarchy, 127, 128
Learning research, 220
Learning styles, 118-120
  assessing, 122
  deductive, 119
  inductive, 119
  levelers, 119
  research evidence about, 119-120
  sharpeners, 119
Learning task analysis, 126-129
  elaboration analysis for information domain, 126-127
  goal analysis for attitude domain, 128
  learning hierarchy for intellectual skills domain, 127
  procedural analysis for motor skills domain, 127-128
Lecture, 141-142
  advantages and disadvantages of, 145-146

lack of interaction a problem with, 142
Legere, C.L., 64
Lessons
 improvement needed in design, 214-215
 sequencing, 129-131
Levelers, 119
Levie, W.H., 5, 134
Linear models, 213
Linstrone, H.A., 53

Maanen, J.V., 191
McCombs, B.L., 240
McCormick, E.J., 84
McKeachie, W.J., 135
McLaughlin, M.W., 10
Macro analysis, 290-292
Macro design, 28, 33-34, 96-104
 illustrated, 96-104
 for Internal Revenue Service *See* Internal Revenue Service macro program
Mager, Robert F., 12, 79, 105
Majority adopters, 183
Management, top down, 240-241
Managers, training needed for, 215-216
Market research, shaping implementation, 211-212
Markle, David G., 16
Martin, B.L., 24
Matching instruction to learners, 219
Mayer, R.E., 134
Mean time between failures (MTBF), 247
Mean time to repair (MTTR), 247
Measurement, distinguished from evaluation, 192-193
Media, 148-152

 computerized, 149-150, 151-152, 154
 factors influencing selection of, 139-140
 matching to learner, 153
 print, 148, 150-151
 transient, 149, 151
 visual aids, 148-149, 151, 153-154
Melching, W.H., 225, 232, 238
Merrill, D.M., 220
Merrill, P.F., 112
Merton, R.F., 112
Messick, S., 119
Methods, instructional, 141-148
 combination approach, 141
 demonstration, 143-144, 147
 discussion, 142-143, 146
 independent study, 143, 146
 lecture, 141-142, 145-146
 on-the-job training, 145, 147-148, 273
 role play, 144-145, 147
 simulation, 144, 147
 tutorial, 143, 146
Methods and media selection, 139-156 *See also* Media; Methods
 based on instructional strategy, 152-153
 cost-effectiveness considerations, 155, 156
 criteria for, 154-156
 cultural factors, 153-154
 feasibility analysis for, 140-141
 importance of visualization, 153
 for Internal Revenue Service micro program, 271-274
 limitations on instructional format affecting, 155

Index

prior familiarity with options affecting, 156
research on effectiveness, 152-154
Micro analysis, 292-294
Micro design, 33, 34, 104-109
 for Internal Revenue Service. See Internal Revenue Service micro program
MicroTICCIT, 226, 239, 242
Military personnel system
 compensatory programs for, 228-229
 function of, 227
 job analysis for, 229-233
 needs analysis of, 227-229
 task analysis for, 229-233
Military personnel training programs
 See also Job Oriented Basic Skills (JOBS) program: Job Skills Education Program (JSEP)
 test Instruments used, 247-248
 time data for, 248
Miller, Robert B., 251
Mogford, B., 225, 232, 238
Motivation, 289-290
Motivation/attitude problems, 74
 solutions to, 78-79
Motor skills domain, 127-128
 procedural analysis of instructional content for, 127-128
 sequencing instructional content for, 130-131

Needs
 interrelationships among, 42
 multidimensional nature of, 41
 origins of, 42-46
 requiring multiple solutions, 45-46, 47
 types of, 41
Needs assessment, 39-64, 292-294
 as a three-step process, 40
 benefits from, 50-51
 data analysis for, 61-64. See also Data analysis
 data collection, 51-56. See also Data collection for needs assessment
 determining training policy, 49-50
 discrepancy model of, 39-41
 importance of, 172
 levels of, 40
 for military personnel system, 227-229
 and organizational environment, 46, 48
 providing baseline performance data, 48
 providing basis for training program, 46-50
 providing description of target audience, 48-49
 providing information on organizational environment, 46, 48
 use of data from, 46-50
Nominal group technique, 53

Objectives, 294-295
 See also Goals and objectives
 changing emphasis on, 219
 conditions component of, 107, 108
 criterion component of, 108-109
 performance component of, 107, 108

Objectives—instructional outcomes, for Internal Revenue Service micro program, 270-271
Observability of innovation, 181-182
Observation
  for data collection, 56, 58
  for task analysis, 85-86
On-the-job training, 145
  advantages and disadvantages of, 147-148
  for Internal Revenue Service micro program, 273
Organizational climate assessment, 54-55
Organizational environment, and needs assessment, 46, 48
Organizational needs, linking training to, 290-292
Organizational problems, 73-74
  solutions to, 77-78
Organizing instructional content, 123-129
  courses, 125
  and domains of learning, 126-129
  learning task analysis for, 126
  lessons, 125
  relationships among programs, curricula, courses and lessons, 123, 124
Ortony, A., 177

Participant analysis, 115-122
  See also Intended participants
  rationale for, 116-124
Patterson, C.H., 176
Performance analysis, 62, 67-79
  and appropriate use of training, 67-68
  consequences of failure to conduct, 69
  examining need for, 69-70
  identifying causes of performance problems, 70-71
  identifying potential solutions to problems, 71-72
  planning for implementation, 72
  procedures for, 69-72
  purpose of, 68-69
  selecting appropriate solutions to problems, 72
  using, 72-79
Performance component of an objective, 107, 108
Performance problems
  exploring solutions to, 75-76
  identifying causes of, 70-71
  identifying potential solutions, 71-72
  isolating causes of, 72-73
  job environment solutions, 77-78
  knowledge/skill factors in, 73
  knowledge/skill solutions, 76-77
  motivation/attitude factors, 74
  motivation/attitude solutions, 78-79
  organizational factors, 73-74
  planning for implementation of solutions, 72
  selecting potential solutions, 72
PERT diagrams, 162, 164-165
Peterson, B.J., 249
Peterson, G.W., 247
Piaget, Jean, 176

Index

Pilot program, 185-187
    for Internal Revenue Service micro program, 280
Pipe, P., 79
PLATO, 226, 239, 242
Poor performance, avoiding rewards for, 79
Pre-experimental designs, 201-203
Prerequisite competencies, 229-232
Prerequisite knowledge, importance of describing, 135
Prerequisites
    converting into learning objectives, 107
    identifying and sequencing, 107
Prerequisite skills, for intellectual skills domain, 127
Print media, 148
    advantages and disadvantages of, 150-151
Problem-solving skills, teaching, 144
Procedural analysis, 127, 128
Process evaluation, 196
Process-helpers, training designers as, 174
Product driven models, 213
Product evaluation, 196
Project agreement, for Internal Revenue Service, 256-258, 283-284
Project management and consulting, 157-170
    designating authority, 158
    estimation for, 159, 160-165. *See also* Estimation
    management skills needed for, 167-170
    personnel involved, 158, 166
    providing a supportive working environment, 167-168
    quality control, 165-166
    reasons for failure, 158-159
    team effort needed, 158, 166-170
Project managers, training needed for, 215-216
Prompts, providing, 136
Provus, M.M., 192

Quality control, 165-166, 214
Quasi-experimental designs, 204-205
Questionnaires, for data collection, 58

Rayner, G.T., 229
Reed, Mildred, 177
Reigeluth, C.M., 90, 126, 129, 133, 135
Relative advantage of innovation, 179-180
Research-based improvements, 241
Resource linkers, training designers as, 174
Retention of instruction, enhanced by visualization, 148, 153
Rice, E.G., 153, 177, 178
Rogers, E.M., 153, 176, 177, 179, 183, 184
Role play, 144-145, 173
    advantages and disadvantages of, 147
Roth, J., 41
Rumelhart, D.E., 177

Salomon, Gabriel, 153
Sample size, 59-60
    and data collection stage, 59-60
    factors to consider, 60
Scheduling, for Internal Revenue Service macro program, 262-263

Schramm, Wilbur, 139-140, 152, 176
Scriven, M., 41
Selectivity in communication, 177
Sequencing of instruction. *See* Instructional content sequencing
Service-wide survey, for Internal Revenue Service micro program, 267
Sharpeners, 119
Shavelson, R.J., 61
Shoemaker, F.F., 176, 183
Showel, M., 225, 232, 238
Simulation, 144, 173
   advantages and disadvantages of, 147
Skill data bases, 218
Smith, M.E., 60
Solution givers, training designers as, 174
Specific entry behaviors, 118
   assessing, 121-122
Stanley, J.C., 201
Stein, F.S., 90
Stitch, T.G., 226
Storytelling research, 177-178
Structured groups, for data collection, 57-58
Structured techniques for reaching consensus, 53
Structure of reality, 177-178
Stufflebeam, D.L., 41, 192, 195, 196
Subject matter experts, use of, 236-237, 238
Summative evaluation, for Job Oriented Basic Skills (JOBS) program, 249-251
Suppes, P., 152
Supportive prerequisites, 107
System-based models, 239

Target audience
   See also Intended participants
   and attractiveness of innovation, 178-182. *See also* Innovation
   cultural bonds affecting, 176-178
   description of, 48-49, 55
   and differences in adoption rates, 182-184. *See also* Adoption
   factors influencing, 176-184
   importance of size of, 110-111
   for military personnel training programs, 245-246
   and needs assessment, 48-49
Task analysis, 81-91
   See also Learning task analysis
   approaches to, 84-88
   by collecting indicator statements, 229
   conducting, 88-90
   contributions of, 83-84
   identifying instructional content, 90
   identifying supporting content for tasks, 90
   illustrated, 12
   for Internal Revenue Service macro program, 258-259
   for Internal Revenue Service micro program, 264-165
   interview data for, 86
   job breakdown for, 84-85
   job task descriptions for, 84, 86
   knowledge or skill analysis for, 87, 90
   for military personnel system, 229-233

Index

observational data for, 85-86
as part of instructional systems development, 82-83
purpose of, 82-83
setting direction of training, 83-84
surveys of job incumbents for, 86
task descriptions information for, 89
task inventory information for, 88
task selection for training for, 84, 86-87
task selection information for, 89
for task sequencing, 84, 87
task sequencing options for, 89-90

Task descriptions, for task analysis, 89
Task inventory, for task analysis, 88
Task selection
    for task analysis, 89
    for training, 84; 86-87
Task sequencing
    for task analysis, 89-90
    for training, 84, 87
Tests, for data collection, 59
Top down management, 240-241
Tracey, W.R., 64
Traditional instruction
    compared with instructional systems development, 16-18, 19-20
    driven by activities during instruction, 104-105
Training
    appropriate use of, 67-68
    linking to organizational needs, 290-291

task analysis setting direction for, 83-84
task selection for, 84, 86-87
task sequencing for, 84, 87

Training design
    cultural factors influencing, 176-178
    differing for large and small projects, 239
    improving on-going training, 214
    likely changes in, 218-220
    for military personnel training programs, 233-236
    research information influencing, 211
    shaped by job design, 212

Training designers
    as catalysts, 174
    as change agents, 172, 175, 184
    as process-helpers, 174
    questions asked by, 172-174
    as resource linkers, 174
    role played by, 174
    as solution givers, 174

Training needs
    identifying, 61-62
    prioritizing, 63-64
Training policies and procedures, determining, 55
Training policy, and needs assessment, 49-50
Training programs, 5
Training strategy, 295-296
Transient media, 149
    advantages and disadvantages of, 151
Trialability of innovation, 181
Turoff, M., 53
Tutorial, 143

advantages and disadvantages of, 146
Tyler, R., 192

Validation stage of data collection, 53-54
Van de Ven, A.H., 53
Visual aids, 148-149
    advantages and disadvantages of, 151
    cultural differences affecting understanding, 153-154
    retention enhanced by, 148, 153

Wager, W.W., 5, 135, 153
Walkthrough, 35
Weaver, R.L., 142
Wedell, E.G., 175
Wells, S., 152
Wilson, L.S., 226
Witkin, B.R., 39
Worth, Sol, 177